Commission
of the European Communities

The book is based on work done in several R&D projects in the framework of an ongoing programme of the COMMISSION OF THE EUROPEAN COMMUNITIES. This programme, called JOULE, supports research and development into renewable energy technologies. It is managed by the Directorate General XII for Science, Research and Development. The synopsis report of three large wind turbine projects, as published in this book, is supported by the Commission under CEC-Contract JOUR-0146-DK.

E. Hau, J. Langenbrinck, W. Palz

WEGA
Large Wind Turbines

With 115 Figures

Springer-Verlag
Berlin Heidelberg GmbH

Dipl.-Ing. Erich Hau
Dipl.-Ing. Jens Langenbrinck

ETAPLAN, Ingenieurbüro für Energietechnische Analysen
und Planung GmbH, München, Germany

Dr. rer. nat. Wolfgang Palz

COMMISSION OF THE EUROPEAN COMMUNITIES,
Directorate General XII for Science, Research and Development

ISBN 978-3-642-52131-7   ISBN 978-3-642-52129-4 (eBook)
DOI 10.1007/978-3-642-52129-4

Typesetting: Camera ready by author

Preface

For almost a decade now, the Commission has been supporting R+D into large wind turbines in Europe within multi-national and pluri-annual programmes in non nuclear energy. The main objective of this action is to further the development of utility-type turbines for mainland electric grid generation and to maintain European leadership in this technology.

The first phase of this action, named WEGA (from the German 'Wind Energie Große Anlagen') has led to the development of three large wind turbines rated at 1.0 to 2.0MW and having rotor sizes, measuring 55 to 60 meters in diameter. The first one, the Tjaereborg 2MW machine, was installed at Tjaereborg near Esbjerg on the west cost of Jutland under the leadership of Elsamprojekt. A short time later, the Spanish-German industrial consortium AWEC-60, joining Union Fenosa with MAN, completed the erection of the second turbine at Cabo Villano, a beautiful cape in the upper western elbow of Spain. Lastly, came the third project, the 1MW machine at Richborough developed by the Scottish firm James Howden with the former UK electricity board CEGB.

These machines have many similarities in their configurations, e.g. they all have three blades and are approximately of the same size, but each of them has different elements of innovation when compared to the previous concepts, either in the electric system, blade design or control philosophy. Also a great effort was devoted to the development of a common data measurement system and programme. For example, the data formats have been fully harmonised and the same software has been running on the three plants to compile periodic operation reports and statistics. Furthermore, all the machines have been instrumented with the same type of equipment, e.g. sensors, strain gauges, that have been mounted at similar locations on the turbine components. By doing so, the errors and distortion imputable to differences in the measurement systems have been prevented. Also the new highly innovative turbine GAMMA 60, which was constructed by WEST in Sardinia, has been equipped with a data measurement system fully compatible with that of WEGA.

Linked to the WEGA programme, a project was therefore set up to define, harmonise, assemble and supervise its measurement programmes. Another important project is underway to gain further experience from the design and operation of these machines. Also the 3-MW machine of Orkney by Wind Energy Group and Scottish Hydro Electric participates in this action. A value-engineering exercise will be carried out on these concepts with a view to single out the most important results and to identify the potential for further design improvement.

During the first year of operation, all three WEGA machines suffered from a major problem which caused some delays in the measurement programmes and additional preoccupations for the project teams. The gearbox of the Tjaereborg machine had to be dismounted when a crack was found in the outer casing. One of the blades of the AWEC-60 had a profound crack and was lifted down and repaired in the nearby workshop. The generator of Richborough machines was removed and repaired due to a earth fault. Now all three machines are operating in good form and are accruing data from the operation and measurement programmes.

Following these activities, the Commission launched a major study involving the best European expertise in large wind energy technology to investigate the potential of advanced large wind turbines to become cost competitive in the medium term. The encouraging results of this work, which will be soon published in the form of a book, prompted the Commission to launch an action to develop a second generation of large wind turbines. This was named WEGA II. While the machines in WEGA were merely experimental machines, this second programme takes into account more the needs of the market and involves major industries in the field. Five completely new prototypes of large horizontal-axis turbines and an other one of a large vertical-axis turbine will be developed in the time frame 1993-1995. From the WEGA I to the WEGA II, an enormous progress has been made in reducing the machine weights and cost. As an example, the blades of this second generation weigh 2 to 4 tonnes vs. 6 to 9 tonnes previously.

In parallel to WEGA II, an extensive R+D programme is being conducted to investigate many unresolved research aspects of large wind turbine technology. This includes projects on aerodynamics, structural analysis, design tools, fatigue, materials and component development. These actions shall provide the technology development for a third generation of large wind turbines (WEGA III) which will eventually constitute the prototypes of commercial large turbines. It must be stressed that for many projects in this programme, WEGA provides the baseline experimental data needed to validate the assumptions made in their models.

This book is an attempt to describe to the general public the WEGA programme and to provide researchers of the field with basic data of scientific and technical relevance. The Commission is very thankful to the authors for having succeeded in this task in a very quick and effective manner. The Commission wishes to congratulate the participants in the WEGA action, and in particular the leaders of the project team - Peter Christiansen from Elsamprojekt, Alfonso Cano and Andres Matas from Union Fenosa, Erich Hau formerly MAN, David Milborrow from National Power and John Rea from PowerGen and their colleagues who have spent an enormous time an effort for the success of WEGA.

Dr. Giancarlo Caratti

Acknowledgements

The authors are very grateful to Mr. Peter Christiansen and Mrs. Peggy Friis from ELSAMPROJEKT, to Mr. Andres Matas from UNION FENOSA, to Mr. Felix Avia and Enrique Soria from CIEMAT-IER, to Mr. Dave Pearce of POWERGEN and to Mr. Peter Simpson from WIND ENERGY GROUP for providing the material of this book. Many thanks also to Prof. Robert Harrison from the University of Sunderland for his assistance with the English translation.

Contents

1

Large Wind Turbine Technology

At the present time, the commercial use of wind energy for electricity production is based on series produced wind energy turbines with power ratings up to approximately 500kW. In a good wind regime turbines of this kind can achieve energy production costs near to conventional generating costs. Very large wind energy converters in the megawatt range have been under development for about 15 years, but as yet their commercial contribution is small.

1.1   Large Experimental Wind Turbines

The development of megawatt wind turbines was initiated mainly by governmental R&D organisations with the intention of providing an additional renewable energy technology which could be used by public utilities. In the United States the first experimental units committed to this aim were erected in 1978. Within a few years a multi-megawatt prototype, the MOD-2, achieved operational status. Five units of this type were built and tested from 1979-82.

In Europe, Sweden and Germany also, showed interest in this technology with the building of experimental machines. The German project, GROWIAN, so far the largest wind turbine ever been built, unfortunately only achieved some hundreds of operational hours, but the large turbines built in Sweden (WTS-3 and WTS-4) have operated for some twenty and ten thousand hours respectively.

Experiments in Canada have had a special focus on vertical axis wind turbines. The world largest Darrieus-rotor, with a capacity of 4MW, has been developed and has been under test since 1987.

Building upon some of the experience gained with this first generation of multi-megawatt experimental wind turbines, a second generation of machines was launched in the mid Eighties. These turbines were somewhat smaller in size - one lesson which was learnt from the first multi-megawatt turbines.

In Britain, Holland and Italy advanced two blade concepts, the LS-1, the NEWECS-45 and the GAMMA 60, have been erected and are being tested.

The Danish utilities, which had accumulated positive experiences from the Nibe experimental turbines, decided to build a 61m diameter, 2 megawatt turbine. This project, later called Tjaereborg Wind Turbine, together with a Spanish-German project AWEC-60 and a British prototype, built by James Howden and Company formed the WEGA large wind turbine programme, which attracted substantial financial support from the Commission of the European Communities.

In total about 25 large wind turbines with power outputs of around one megawatt and more have been erected and evaluated in the period from 1978 to 1993. In general terms it can be concluded, that nearly all of these turbines have successful demonstrated their functional integrity from the technical point of view. All projects have proved to have had a larger R&D component than was generally expected in the beginning of the development and there have been a number of technical problems and failures.

The technical problems have either been solved or are capable of solution within a short time scale, but the main obstacle to the commercial use of the large wind turbines remains inferior economics. The specific installation costs (ECU/kW or ECU/m²) for the machines are up to four times higher than those for small and medium sized turbines. The higher energy captures resulting from the utilisation of the higher wind speeds obtained at the higher rotor hubs is not sufficient to compensate for this effect. As a result the economics of the large wind turbines are remarkable inferior to the smaller commercial machines.

The main reason for the high production costs of the megawatt wind turbines lies in the high component weights. A comparison with the today's smaller commercial turbines highlights the unfavourable specific towerhead weights of the larger machines (Tab.1.1). Although there are intrinsic physical reasons for an increase of the specific weights per swept rotor area, a significant potential for weight and cost reduction can still be identified.

MOD-2 (USA), 91mØ, 2.5MW, 1979

WTS-4 (USA), 78mØ, 4MW, 1981

WTS-75 (S), 75mØ, 2MW, 1983

GROWIAN (D), 100mØ, 3MW, 1984

Aéole (CAN), 68mØ, 4MW, 1987

Tjaereborg (DK), 61mØ, 2MW, 1988

NEWECS-45 (NL), 45mØ, 1MW, 1985

MONOPTEROS (D), 56mØ, 640kW, 1990

	Commercial Wind Turbine	3-Bladed EC-R&D Turbines WEGA Programme			2-Bladed EC-Demo Turbines			1-Bladed Turbines
	Vestas V25 DK	Tjaereborg WT DK	Richborough WT UK	AWEC-60 E	GAMMA 60 Alta Nurra I	NEWEC 45 Wieringer NL	HMZ 1 MW Zeebrugge B	MONOPTEROS Wilhelmshaven D
Rotor Diameter [m]	25.0	60.0	55.0	60.0	60.0	45.0	45.0	56.0
Year of Completion		1988	1989	1989	1990	1985	1990	1989
Towerhead Weight [tons]	10.3	224.4	83.4	183.5	95.0	73.0	98.0	61.2
Total Weight [tons] (incl. tower, excl. foundation)	21.3	887.4	178.0	275.5	245.1	153.0	193.0	146.2
Swept Area [m²]	491	2,827	2,375	2,827	2,827	1,590	1,590	2,463
Spec. Towerhead Weight [kg/m²]	21.0	78.7	35.1	64.9	33.6	45.9	61.6	24.8
Rated Power [kW]	200	2,000	1,000	1,200	1,500	1,000	1,000	640
Spec. Towerhead Weight [kg/kW]	51.5	111.2	83.4	152.9	63.3	73.0	98.0	95.6
Energy Production [kWh/m²/y]	1,000-1,340	1,380	910	1,110	1,075	1,380	880	860
Spec. Towerhead Weight [kg/kWh/y]	0.015-0.021 (1)	0.057 (0.057) (2)	0.039 (0.033) (2)	0.057 (0.062) (2)	0.031 (0.026) (2)	0.033 (0.033) (2)	0.043 (0.075) (2)	0.029 (0.027) (2)

(1) Typical range based on reported experience

(2) Data given as values predicted for actual site (corrected to a site of annual mean wind speed = 7.7m/s at 60m height). Assumed availability 95 %.

Tab.1.1 Specific Weight and Performance Data of Large New Wind Turbines in the EC

The technical concepts of the today's experimental large turbines have to be considerably improved in order to give more lightweight and cost effective designs. To achieve this goal continuous R&D activities are necessary leading towards the next generation of large wind turbines. The WEGA large wind turbine programme is committed to this aim.

1.2   Two Blade Concepts: LS-1 and GAMMA 60

Two European large wind turbines, the LS-1 and the GAMMA 60, feature two bladed rotors. The three WEGA wind turbines all use the more conventional three bladed design and hence a comparison with the two bladed technical line is of special interest. Against this background, and with the financial support of DG XII of the Commission of the European Communities the Italian GAMMA 60 has been included in the WEGA measurement project, and the British designers of the LS-1 are participating in the WEGA design review and evaluation. This is being carried out on the basis of pre-existing measurements within the framework of the British national programme.

LS-1 3MW Wind Turbine

The LS-1 is a 60m diameter, horizontal axis wind turbine rated at 3MW. It is located at a site on Burgar Hill on Mainland Orkney, the largest in the group of islands situated immediately to the north of Scotland (Fig.1.2). The installation has been designed and built by the Wind Energy Group (WEG), a joint venture of Taylor Woodrow Construction, British Aerospace and GEC Energy Systems Ltd., under contract to the North of Scotland Hydro Electric Board and the UK Department of Energy (now the Department of Trade end Industry). The general arrangement of the plant at the top of the tower is shown schematically in Fig.1.3., and can be briefly described as follows:

Rotor
The rotor has two blades with a teetered hub, mounted on elastomeric bearings. The outer 30% (9m length) of the blades is mounted on rolling element bearings, permitting variation of pitch setting. A GFRP spinner is mounted on the rotor hub and provides an enclosure for the hydraulic systems controlling tip blade pitch settings and teeter restraint.

Drive Train
The primary, 2-stage epicyclic gearbox with mechanical brake, is supported on rear trunnion mountings located at the level of the point of intersection of the longitudinal axis of the gearbox with the centreline of the tower, and on strut supports located beneath the front bearing carrying the main shaft. A bevel gear with a vertical output drive down the axis of tower provides the third stage. A secondary, single stage epicyclic gearbox is mounted on the generator, located at the top of the tower, supporting a variable speed synchronous reaction machine, engaging with the sun gear to provide control of the speed of the rotor, within a range of ±5% of the nominal rpm.

Fig.1.1 LS-1 Wind Turbine at Burgar Hill (Orkney Islands)   (Photo WEG)

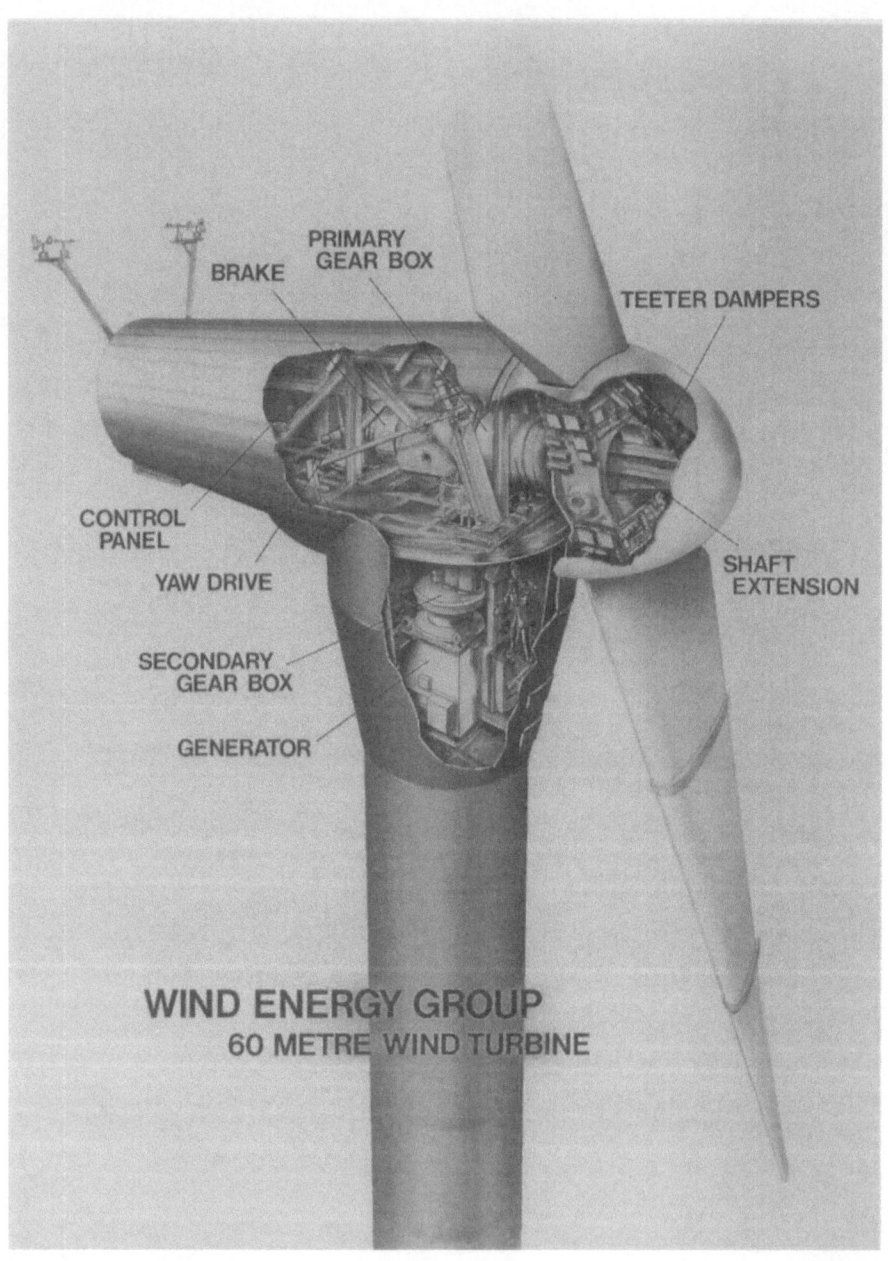

Fig.1.2 Technical Arrangement of the LS-1 Wind Turbine   (Source WEG)

Rotor	
	Number of Blades 2
	Height to Rotor Hub 45m
	Diameter 60m
	Rated Speed 34rpm
	Tilt Angle .... 5°

Performance	
	Rated Power 3MW
	Wind Speeds (Hub Height):
	- cut-in 7m/s
	- rated 17m/s
	- cut-out 27m/s
	- survival .... 70m/s 3-sec gust
	Power Coefficient .... $c_{p,max} = 0.43$
	Tip Speed Ratio at $c_{p,max}$ 8

Blades	
	Material .... steel + GFRP
	Aerofoil .... NACA 44XX series
	Control .... pitch of outer 30% span
	Pitch Mechanism .... hydraulic

Hub	
	Type .... teetered
	Teeter Angle .... +5°

Transmission	
	Primary Gearbox .... 2 stage epicyclic + bevel
	Ratio 31.3
	Brake .... calliper/6 disc

Generator	
	Type .... synchronous
	Rated Power 3,750kVA
	Voltage 11kV
	Speed 1,500rpm

Yaw Drive	
	Type .... electric motors
	Rotation Rate .... 10°/minute

Tower	
	Type .... prestressed concrete
	Diameter of Cylinder 3.8m

Fig.1.3 Technical Specifications of the LS-1 Wind Turbine

Nacelle

The nacelle consists of a structural frame of as-rolled steel members, supporting the primary gearbox; a derrick for lifting and lowering the complete rotor plus spinner assembly and (separately) the primary gearbox; ancillary plant; and sheeting rails carrying aluminium alloy plain sheet cladding.

Yaw Mechanism

The yaw mechanism consists of a bearing comprising four pads of phenolic resin impregnated asbestos fabric with a filled PTFE surface layer, carrying the vertical load and sliding on a machined steel ring welded to the tower head. There are three centralising wheels, mounted in the nacelle, driving a pinion engaging with steel pins carried in the main bearing ring.

Electrical System

A synchronous generator is mounted in the top section of the tower in an arrangement that permits removal and replacement, with the aid of purpose built equipment and the ground mounted winch that serves the nacelle mounted derrick.

Tower

The tower has a 37m high prestressed concrete lower section, comprising a 15m high base frustum and cylindrical shaft, with a 2.8m deep octagonal plain reinforced concrete foundation. A reinforced concrete slab at the 4m level isolates plant room at ground floor level from the main tower section. The top section is fabricated in steel in the form of an inverted frustum, 6m high terminating in the 5.8m diameter yaw mechanism main bearing ring; with an aluminium alloy plain sheet cladding, matching the nacelle. The tower is dynamically 'stiff' i.e. the first mode natural frequency of the complete installation is above the blade passing frequency (1.13Hz).

Operational Record

Since the machine was commissioned on 20.11.92, it had accumulated 6,846 hours of running time and produced 9,204MWh of electricity.

A 12 month monitoring period was initiated on 28.2.92. Since that time, and up to 20.11.92, the performance has been as follows:

Running Time	2,591 hours
Energy Production	3,420 MWh
Availability	68.8 %
Load Factor	17.9 %

GAMMA 60 1.5MW Wind Turbine

The GAMMA 60, a 2 bladed, 1,500kW rated power wind turbine, has been designed by AERITALIA under joint contract with ENEA and ENEL and with financial contribution from the Commission of the European Communities, Directorate General for Energy. The design characteristics of the GAMMA 60 were conceived specially to meet the requirements of the Italian electrical utilities.

Design Features
The advanced concepts of the GAMMA 60 include two main innovative elements:

- Yaw drive system utilised for power regulation.
  This procedure enables the blade pitch system to be eliminated, thus reducing the costs of many components in the machine.

- Operation in broad range variable speed mode, to increase the energy capture and the fatigue life.

Rotor
The rotor consists of a steel hub with 2 blades made of a fibreglass and resin composite material. Each blade is 29m long, which gives a rotor diameter of 60m when the hub is included. Elastomeric bearings and bumpers in the hub enable the rotor to teeter.

Drive Train and Nacelle
The main shaft consists of a single block of forged steel supported by rolling bearings set in a very strong housing. It is hollow to allow signal wires to pass through. The rotor torque is transmitted to the electric generator through the main shaft, the gear box and the high speed shaft. The yaw drive device includes a bearing support with two yaw motors.

The Electrical and Control System
The electrical subsystem includes a synchronous generator at variable frequency and voltage. A static voltage and frequency converter is used to allow connection to the grid. The command and monitoring system ensures that control is maintained both in normal and in emergency operating modes.

Tower
The tower is a free standing steel tube design. Inside the tower, power and signal wires are located. A lift provides access to the nacelle.

Fig.1.4 GAMMA 60 1.5MW Wind Turbine at Alta Nurra (Sardinia)   (Photo ENEL)

Rotor	Number of Blades	2
	Diameter	60m
	Height to Rotor Hub	66m
	Rated Speed	35.5...44rpm
	Cone Angle	3°
	Tilt Angle	6°
	Power Control	yaw control
Performance	Rated Power	1,500kW
	Wind Speeds (Hub Height):	
	- cut-in	5m/s
	- rated	13.3m/s
	- cut-out	27m/s
	- survival	64m/s
Blades	Material	GFRP
	Aerofoil	NACA 230XX series
	Chord Tip / Root	1.06m / 3.82m
Hub	Type	teetered
	Material	casted steel
	Diameter	2.54m
Gearbox	Type	2 stage epicyclic
	Ratio	1:33
Brake	Type	emergency disk brake
	Position	main shaft
Generator	Type	synchronous
	Rated Power	2,000kW
	Voltage	12kV
	Speed	variable
Yaw Drive	Type	hydraulic, redundant
	Rotation Rate	2°/s...8°/s
Tower	Type	tubular, cantilever, steel
	Diameter	3.0m

Fig.1.5 Technical Specification of GAMMA 60

2
Outline of the WEGA Large Wind Turbine Programme

Within the context of the continuing development of large wind turbine technology the WEGA turbines can be regarded as the 'second generation' of large experimental machines. Designed in the years 1984 to 1986, they built upon experience gained from the 'first generation' of large experimental machines which were erected in the early eighties. The MOD-1 and MOD-2 (USA), GROWIAN (DE) and Nibe (DK) are examples of these. The common intention of the designers of the WEGA turbines was to use the lessons learnt from the building and operation of these first large machines but to combine this with the experience which had emerged from the successful development of the first small commercial wind turbines. In this way the technology would proceed one step further in the process of the commercialisation of wind turbines in the megawatt class.

Nevertheless it was clear from the outset that the development of wind turbines in this size range would require large extended research projects. This has become even more obvious as the work programmes have proceeded. It is a common feature of the development of the three machines that although they were constructed by quite different groups of companies with different underlying design philosophies, a strong engineering research environment was available to support the design effort in each case.

2.1   Technical Background and Design Philosophy of the WEGA Turbines

Tjaereborg 2MW Wind Turbine

In Denmark cooperation between the Danish Ministry of Energy and the electric utilities, in the field of wind energy, started in 1976 with a joint programme directed towards the development of large wind turbines, and the investigation of the problems of siting and their integration into the electric grid. One of the activities in the first years of this programme was the design, construction and commissioning of the two NIBE Wind Turbines. With a rotor diameter of 40m and a rated power of 630kW they were at that time among the largest in the world. As no main supplier, which could offer a turn key responsibility, was available, the turbines were built using the multi contract approach which is normally used by the electrical utilities in Denmark when building large power plants. The commissioning and first phases of the test period of the two turbines were finalised during the early part of 1981.

Fig.2.1 NIBE Wind Turbines, 1981, Rotor Diameter 40m, Rated Power 640kW

The results with the NIBE turbines were positive and a new project, designated the 'K-Project', with the goal of designing a new, larger wind turbine utilising the experience gained with the NIBE turbines was formulated and started in the summer of 1981 (Fig.2.2). This used the successful original team which comprised the design staffs of the utilities ELKRAFT and ELSAM together with aerodynamic specialists of the Department of Fluid Mechanics at the Technical University of Denmark.

During the first phases of this project the pitch controlled NIBE B turbine was chosen as the model for the new design, rather than the stall controlled NIBE A. This was partly due to the unresolved problem with the stall induced vibrations in the NIBE A turbine in the high wind speed range, and partly due to the better start-up and stop characteristics of the pitch controlled machine. The diameter of the new design was primarily decided from an evaluation of the extent which the GRP blade technology of the outer 12m of the NIBE blades could be scaled up in size.

Fig.2.2 'Project K', 1983, Rotor Diameter 60m, Rated Power 2MW

The result of the project was a technical description of a pitch controlled wind turbine with a rotor diameter of 60m, a hub height of 60m and with a 2MW induction generator. Simultaneously with this, a 30m blade was developed and a test blade was produced and tested.

In 1984 it was found desirable to divide the tasks of the Danish wind energy programme between the two utility organisations, ELKRAFT and ELSAM, and it was decided that ELKRAFT should concentrate upon the Masnedø Wind Farm and that ELSAM should continue with the 2MW wind turbine. The detailed design work was started in order to establish a cost basis for the construction decision and in 1985 ELSAM decided to build the turbine provided that supplementary funding could be found.

The contract with the European Commission, DG XII was signed on December 1, 1986 and the work with the detailed design of the turbine and its construction commenced. The utility I/S Vestkraft was appointed as principal on behalf of ELSAM and acts as the operating agent of the turbine today.

The design which arose out of the 'Project K', later named the 'Tjaereborg Wind Turbine', was derived from the Nibe B wind turbine, i.e. a three-blade upwind rotor on a rigid hub, a traditional drive train with an induction enlarged-slip generator, power regulation by electrohydraulic full-span pitch control and a low-tuned concrete tower. This tower concept is still very novel and its suitability for larger machines has not yet been established conclusively.

The technical aims of the Tjaereborg project were concentrated on the following items:

- to design and test a 30 metre blade suitable for series production.
- to construct a 2MW turbine of a simple and robust design for a high availability,
- to operate the turbine connected to the common electrical grid and to determine the costs of operation and maintenance,
- to perform measurements, analysis and design verifications,
- to evaluate the results and thus improve future designs so that they have potential for low-cost production.

A very special factor in this project has been the heavy involvement from the electrical utilities.

Richborough 1MW Wind Turbine

The technical background of the Richborough wind turbine differs considerably from the Tjæreborg or the AWEC-60 development. The 1MW wind turbine at Richborough/Kent in the United Kingdom has been developed in the framework of an industrial product line.

The James Howden &Co. of Glasgow started with the development of wind turbines in the early Eighties. In 1983 the first 300kW wind turbine with a rotor diameter of 22m was erected at the test field Burgar Hill on the Orkney Islands. The wind turbine was owned by the North of Scotland Hydro Electric Board. The HWP-300 featured a three blade rotor with movable blade tips for speed and power control. An innovative wood/epoxy material was used for the rotor blades. This wind turbine has proved to be very servicable in the harsh wind regime of the Orkney Islands.

Fig.2.3 HWP-300 Wind Turbine at Burgar Hill, Orkney Islands, 1983

Based on the success of this test machine James Howden &Co. subsequently developed a commercial version of the same design. This machine could be fitted with different rotors of size from 26m to 31m giving rated powers of up to 330kW. In 1984 a 26MW wind farm at Altamont Pass in California was built with 75 of these units. In the beginning of operation the machines suffered a series of blade failures, with the result that the manufacturer had to retrofit all rotors with an improved blade root attachment.

Fig.2.4 Wind Farm at Altamont Pass/California Consisting of 75 Howden HWP-330 Wind Turbines, 1984

The next stage in the process of increasing the size of the same technical concept was the development of a prototype machine with a 45m rotor and a rated power of 750kW. The first test machine of this class was erected at the Altamont Pass wind farm site in 1984. Two other units of this type were also delivered, one to Sweden and the other to the North of Scotland Hydro Electric Board to be erected on the Shetland Islands.

Fig.2.5 Howden HWP-750 Wind Turbine, Rotor Diameter 45m, Rated Power 750kW, 1987

To move beyond this size, in order to meet the requirements of the utilities for megawatt scale wind turbines, a 1MW wind turbine project with a rotor diameter of 55m has been created. The former national electricity board the CEGB selected a site near the Richborough power station in Kent for this project.

By the end of 1986, having successfully applied for financial support, the CEGB signed contracts with the European Commission, the UK Department of Energy and James Howden &Co. for the development of the Richborough wind turbine as part of the WEGA programme.

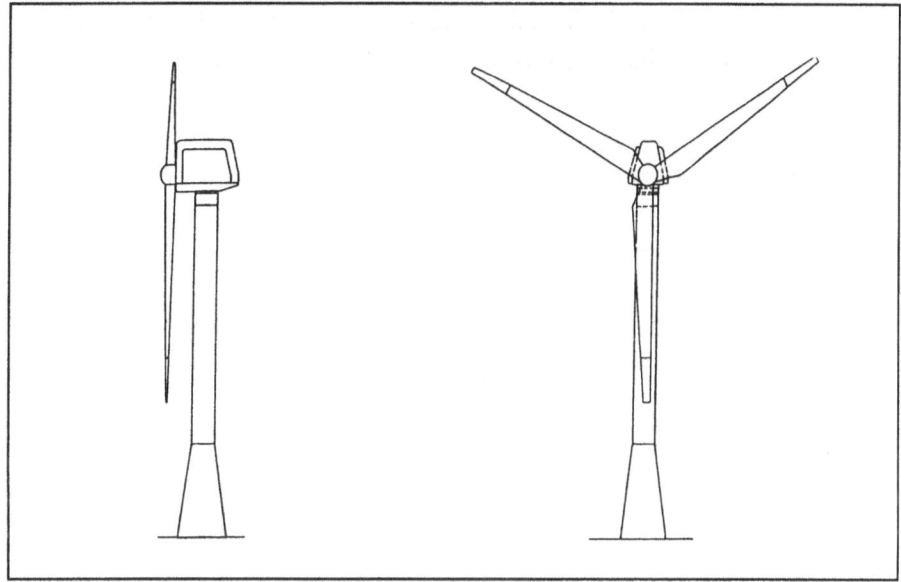

Fig.2.6 Richborough Wind Turbine Project, 1986, Rotor Diameter 55m, Rated Power 1000kW

The technical aims of the project were:

- the development of the lightweight wooden rotor incorporating tip speed control up to the megawatt range

- to extend the proven conceptual design of the former 330 - 750 kW units up to a megawatt turbine.

The Richborough wind turbine can be seen as an early industrial attempt to develop a commercially viable megawatt wind turbine.

AWEC-60 1.2MW Wind Turbine

The AWEC-60 concept has developed as a result of Spanish-German co-operation in this field, but it was based upon earlier experience obtained from large wind turbine developments in Germany.

In the late seventies in Germany the large experimental wind turbine, project GROWIAN, was initiated by the German Ministry for Research and Development (BMFT). The main industrial contractor was the MAN-Maschinenfabrik Augsburg Nürnberg AG in Munich. The 3MW GROWIAN was erected in 1981 and tested over a three year period. The test unit did not achieve continuous operation due to recurrent, severe, material fatigue damages.

The results from this experiment and from similar projects in USA and in Sweden showed that wind turbines in the multi-megawatt power range are far from being commercially viable. However, they did demonstrate the technical feasibility of very large wind turbines.

Against this background a new project, named WKA-60 was launched in 1982.

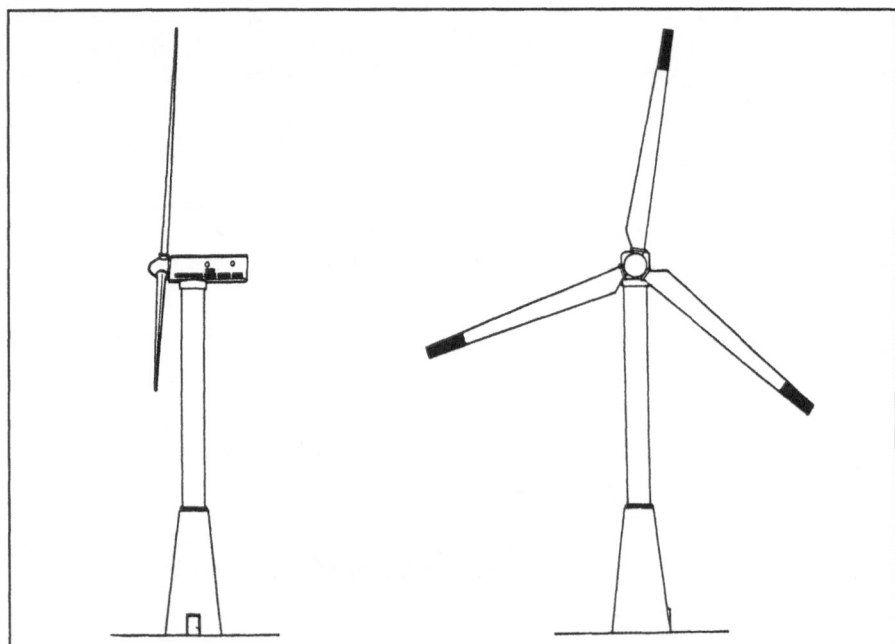

Fig.2.7 Project WKA-60, 1982, Rotor Diameter 56-60m, Rated Power 1200kW

The technical concept of the WKA-60 was permeated by the basic idea of producing a medium - large wind turbine with low development costs and easy operating and maintenance characteristics. The technical design approach was rather conventional, in comparison with GROWIAN, or i.e. with the single bladed MONOPTEROS wind turbine, which has been tested in Germany at the same time. A three bladed rotor was chosen in order to avoid the dynamic problems of the two or one blade rotors. The mechanical drive train was designed to be modular so that it could be based, as far as possible, on series produced components.

The WKA-60 concept formed the technical basis for three different units:

One test unit was designed for a site on the German Island of Heligoland. This project was supported by the German Ministry for Research and Technology (BMFT) and the Commission of the European Communities (CEC) within the frame work of the wind energy demonstration programme.

Fig.2.8 WKA-60 Wind Turbine on the German Island of Heligoland, 1990 Rotor-Ø 60m, Rated Power 1200kW

The second unit, named AWEC-60 (Advanced Wind Energy Converter - 60), was developed through a Spanish-German cooperation within the framework of the WEGA-programme. This version incorporated some innovative elements which determined the research character of the project:

- an innovative variable speed electrical generator system with the built-in capability to operate the wind turbine in three different speed and control modes

- a new Spanish concept of rotor blade design and manufacturing

The aim of the project was to carry out an experimental research programme concentrating on these innovative components and to introduce large wind turbine technology into the Spanish wind energy programme in general.

In 1985 the AWEC-60 project was conceived by the Spanish-German Consortium formed by MAN NEUE TECHNOLOGIE GmbH, two Spanish research organisations CIEMAT-IER and ASINEL and the utility UNION ELECTRICA FENOSA.

It should be mentioned that a third machine in the WKA-60/AWEC-60 series was manufactured in 1991. This test unit is a derivative of the Heligoland wind turbine incorporating some cost reducing elements, i.e. a prefabricated concrete tower. The turbine was erected at the former GROWIAN site (Kaiser-Wilhelm-Koog) and will be operated by the German utility Preußen-Elektra.

Fig.2.9 WKA-60 II at Kaiser-Wilhelm-Koog, 1992

2.2   Sites and Wind Resources

The sites of the three WEGA turbines are located in the western coastal area of Europe and exploit the good wind resources which exist along the shores of the Atlantic Ocean and of the North Sea (Fig.2.10).

The Tjaereborg site lies within a "classic" Danish wind energy area whereas the use of wind energy at Richborough, in the south-east of England, is rather new. The third WEGA turbine, the AWEC-60, is located in the North West corner of Spain, an area which is exposed to high onshore winds from the Atlantic ocean.

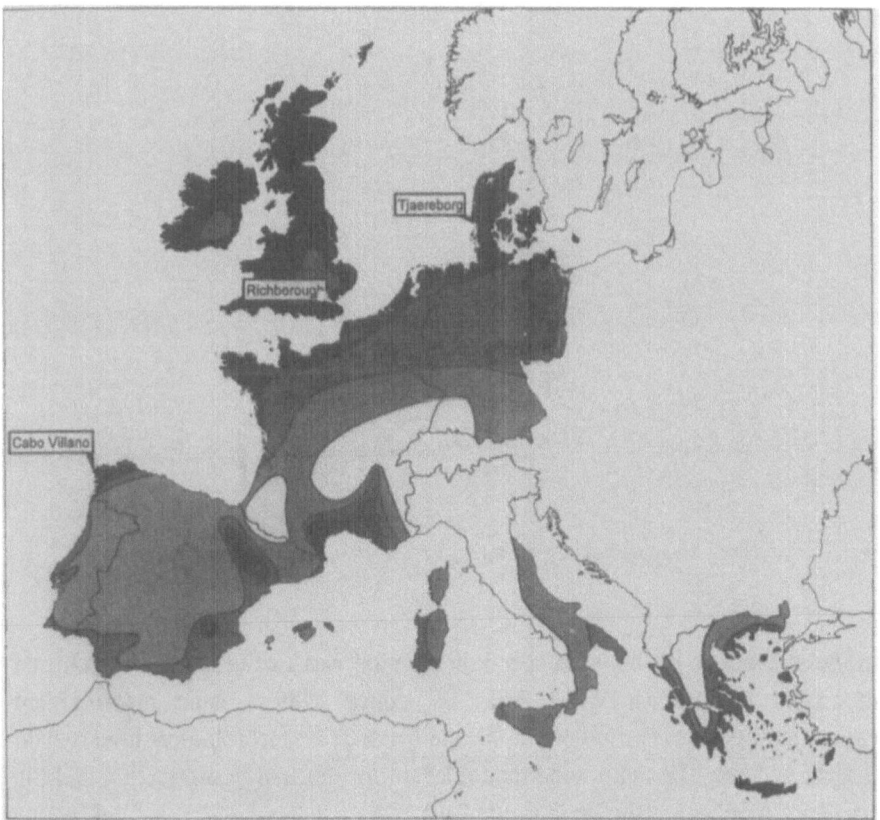

Fig.2.10 Sites and Wind Resources of the WEGA Turbines Against the Background of Wind Speeds in Western Europe (taken from the European Wind Atlas /1/.)

Tjaereborg

The Tjaereborg 2MW turbine location is in the western part of Denmark. The site is just south-east of Esbjerg (about 9km) in meadows between a small village (Tjaereborg) and the coast (about 800m) as shown on the site plan (Fig.5.10).

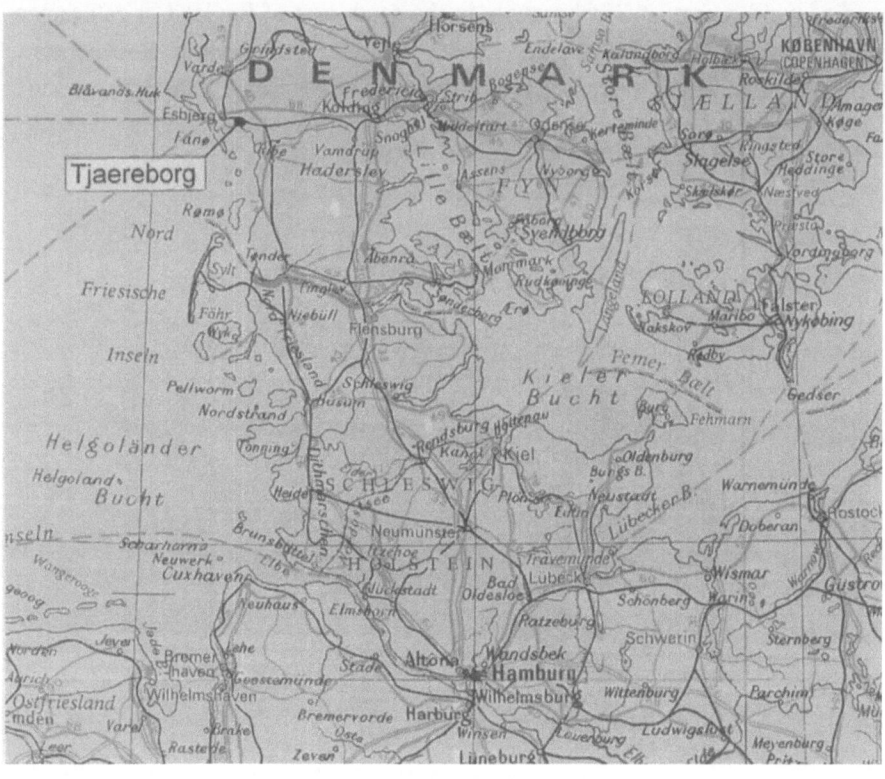

Fig.2.11 Tjaereborg Site near Esbjerg (Denmark)

An initial estimate of the wind speed distribution was carried out for the chosen site, using the Danish Wind Atlas /2/. Since 1985 a wind measurement programme has been carried out at the site using two meteorological towers with a height of 50m. The main wind data relevant for the energy production and the turbulence loading are shown in figure 2.12.

Mean Wind Speed at 10m Height:....................6.9m/s
Mean Wind Speed at Hub Height, 60m:............7.7m/s
Mean Vertical Wind Shear Exponent:...............0.06
Wind Speed Distribution:
(Weibull parameters)....................................A = 8.65m/s
C = 1.76
Mean Turbulence Intensity (above 8m/s):..........≈ 10%

Fig.2.12 Main Wind Data at Tjaereborg

Richborough

The location of the Richborough 1MW wind turbine is at the Richborough Power Station Site in Kent owned by PowerGen. The site is 4.5km north of Sandwich, 2km west of the mouth of the River Stour and approximately 5km south-west of Ramsgate. The machine is located to the north-west of the existing power station and is approximately 5m above sea level. The surrounding terrain is mostly level, but the power station presents a fairly large obstacle, with 100m high cooling towers lying at a distance of 1km from the machine.

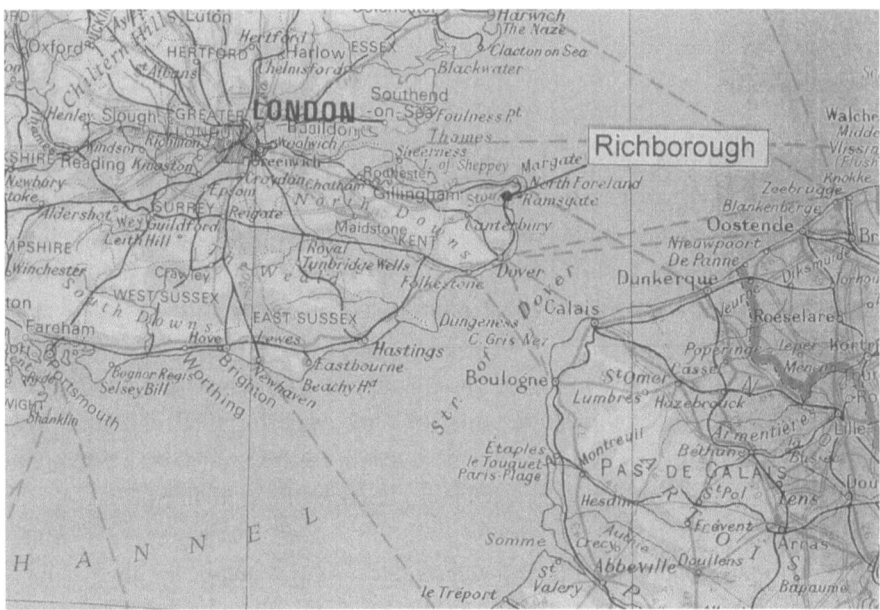

Fig.2.13 Richborough Site in Kent, United Kingdom

The wind conditions have been measured at Richborough since 1981 in anticipation of the siting of a new large wind turbine. The measurements were made using a lighting tower, very close to the proposed wind turbine site. They have been made at heights of 10, 22 and 49m. Comparisons of measured estimates of mean wind speed with equivalent data from a nearby meteorological station, 3km to the NE have also been made, to enable the long term average to be predicted. Additional comparisons have been made with wind speed predictions derived from upper air data. Satisfactory agreement between the techniques was obtained, reflecting the good exposure and low roughness of the chosen site (Fig.2.14).

Mean Wind Speed at 10m Height:..................... 4.82m/s
Mean Wind Speed at Hub Height, 60m:............ 6.83m/s
Mean Vertical Wind Shear Exponent:............... 0.23
Wind Speed Distribution:
(Weibull parameters).. A = 7.99m/s
                                                                      C = 2.18
Mean Turbulence Intensity (above 8m/s):.......... 12%

Fig.2.14 Main Wind Data at Richborough

Cabo Villano

The AWEC-60 machine is located on the Cabo Villano peninsula in the municipality of Camarinas, province of La Coruna (Spain). The site is about 1km from the Cabo Villano lighthouse, a very well known point for sailors coming from the south to the Bay of Biscay. The AWEC-60 is installed on a point 81m above sea level at a distance of about 220m from the seashore.

In July 1985, a 10m meteorological mast equipped with sensors for wind speed, wind direction, air temperature and pressure was installed on site. This meteorological station was equipped with a data recorder and pre-processing unit which gathers values of variables each hour (or each fifteen minutes).

In May 1986, an additional 50m meteorological mast, equipped with four level sensors for wind speed, two level sensors for wind direction and sensors for temperature, atmospheric pressure and humidity was erected at the site.

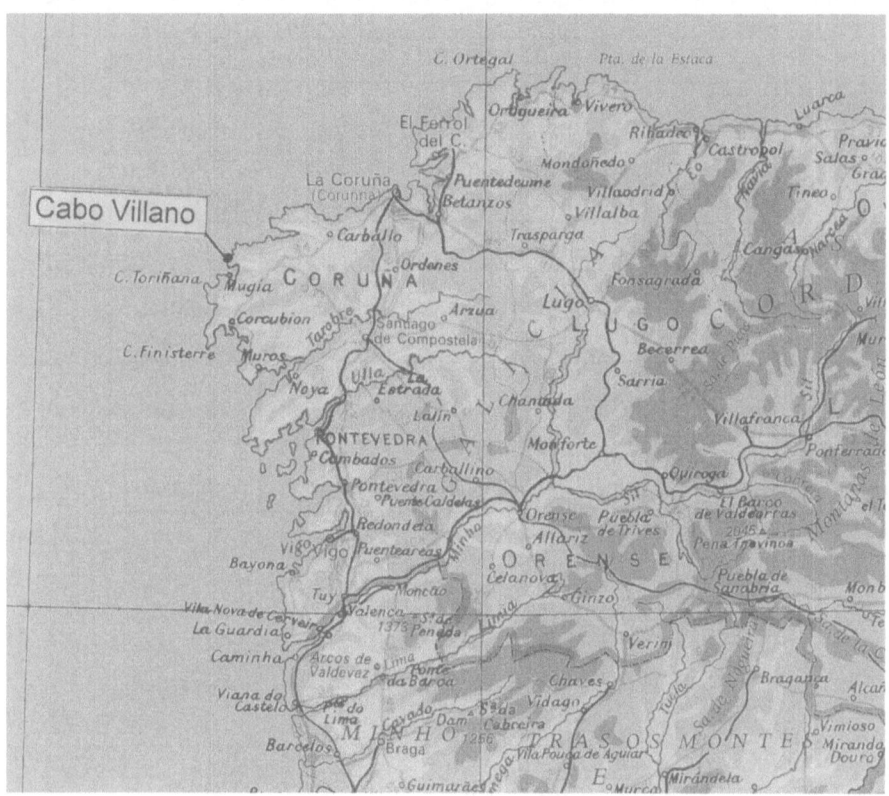

Fig.2.15 Cabo Villano in the North West of Spain.

The processing of data was carried out in accordance with the measurement programme prepared in July 29, 1986 ('Wind Measurement Programme for evaluation of Cabo Villano site'). The main wind data of Cabo Villano are summerized in Fig.2.16.

Mean Wind Speed at 10m Height:......................6.10m/s
Mean Wind Speed at Hub Height, 60m:.............7.70m/s
Mean Vertical Wind Shear Exponent:................0.09...0.15
Wind Speed Distribution:
(Weibull parameters)...A = 8.24m/s
                                                                              C = 1.95
Mean Turbulence Intensity (above 8m/s):...........≈ 6%

Fig.2.16 Main Wind Data at Cabo Villano

A summary of the wind data at the WEGA sites is given in the following table.

	Tjaereborg	Richborough	Cabo Villano
Mean Wind Speeds			
at 10m height	6.90m/s	4.82m/s	6.10m/s
at hub height	7.70m/s	6.83m/s	7.70m/s
Shear Wind Exponent	0.06	0.23	0,09...0,15
Wind Speed Distribution			
Weibull Parameters			
A =	8.65m/s	7.99m/s	8.24m/s
C =	1.76	2.18	1.95
Mean Turbulence Level (at wind speeds above 8m/s)	appr. 10%	appr. 12%	appr. 6%
Predicted Energy Production (assuming 100% availability)	4.1GWh	2.4GWh	3.5GWh

Fig.2.17 Summary of the Wind Data at the WEGA Sites

2.3   Project Organisation, Budgets and Time Schedule

The design, construction and operation of large wind turbines is dominated by the R&D component of the projects. While industrial manufacturers and their scientific and engineering sub-contractors carry out most of the work, the main purpose of the development is to satisfy the needs of the electrical utilities. The organisational structure of the WEGA wind turbine programme has been tailored to this, with the involvement of utilities being a major aspect. Close cooperation between utility and manufacturer has been found to be an important feature contributing to the success of each of these projects.

Tjaereborg Wind Turbine

The Tjaereborg wind turbine is owned by the Danish utility Vestkraft which is based in Esbjerg. Construction, erection and commissioning have been carried out in a combined working team, consisting of members from Vestkraft, The Technical University, Elsamprojekt, and parts of the Danish Industry. Vestkraft and Elsamprojekt are members of the Danish utility cooperation, ELSAM I/S.

In contrast with the other WEGA wind turbines no industrial company or consortium served as main designer nor as main contractor. In accordance with the Danish tradition and, the experiences gained with the Nibe wind turbines, ELSAMPROJEKT A/S performed the design work and system level work as well as providing management during the construction phase.

The design and construction costs for the wind turbine are summarised in the table below. The total cost of 9.35 MECU implies a 13% overrun on the inflation corrected original budget. The cost breakdown in the main areas is:

Design and administration:	37.8%
Grid connection:	1.5%
Electrical systems:	2.9%
Control system:	2.2%
Mechanical systems:	37.2%
Auxiliary systems:	1.3%
Buildings, including tower:	10.3%
Measurement system:	6.8%

Financing was obtained from three sources, The Danish Ministry of Energy, The Commission of the European Communities, DG-XII and ELSAM:

European Commission:	2,100,000 ECU
The Ministry of Energy:	
Design ,Construction and	
Measurement programme:	2,575,000 ECU
ELSAM:	4,680,800 ECU
Total	**9,355,800 ECU**

Richborough Wind Turbine

The design and construction of the Richborough wind turbine has been undertaken by a joint venture between James Howden and Company Limited of Glasgow and the former utility the Central Electricity Generating Board (CEGB). In the course of the reorganisation of the british power generating and distribution industry, PowerGen, as one of the successors to the CEGB has inherited the Richborough power plant and has operated the wind turbine since 1990. The manufacturer James Howden &Co., has now abandoned wind energy activities but has transferred the technology to the company Windharvester Ltd..

The total budget of the project has been calculated in 1986 to be:

£ 3,454,000 = ECU 4,446,900

The costs have been shared by:

European Commission:	1,528,900 ECU
UK Department of Energy:	1,097,900 ECU
CEGB/PowerGen:	1,528,900 ECU
James Howden:	291,200 ECU

AWEC-60

The AWEC-60 wind turbine has been developed and constructed by a consortium consisting of:

UNION ELECTRICA FENOSA
ASINEL
CIEMAT-IER
MAN, Maschinenfabrik Augsburg Nürnberg AG

Because the AWEC-60 wind turbine is basically the same design as the WKA-60 on the island Heligoland it was possible to use some identical components in the design. In terms of organisation the AWEC-60 project team subcontracted to MAN Technologie the supply of the drive train and pitch control of the blades. Then in order to achieve some cost reduction and to increase the cooperation between both Spanish and German partners, MAN has subcontracted on to the Spanish participants the supply of the nacelle and yaw drive. MAN Technologie AG has carried out all the engineering of the mechanical parts and the Spanish partners undertook the design of the electrical system and rotor blades.

The budget of the project has been calculated in 1986 to be 13,960,000 ECU. Financing was shared by the following sources:

European Commission:	2,130,000 ECU
BMFT, Bundesministerium für Forschung und Technologie (to MAN)	3,500,000 ECU
CIEMAT-IER	1,760,000 ECU
ASINEL	310,000 ECU
UNION ELECTRICA FENOSA	6,260,000 ECU

Time Schedule

The three WEGA wind turbines were developed and constructed during the period from 1985-1989, while the first pre-design studies, i.e. for the Tjaereborg wind turbine, date back to the year 1982. A data acquisition and evaluation period of '12 month of availability' was foreseen in the original contracts. However, due to long interruptions in the test phases of all three projects, a considerably longer time has been needed for the data acquisition. To ensure a comprehensive data exchange and common evaluation two additional scientific projects have been established. The WEGA 'Measurement Project' and the 'Design Review and Evaluation' are both being carried out by ELSAMPROJEKT as coordinator and will continue until the end of 1993.

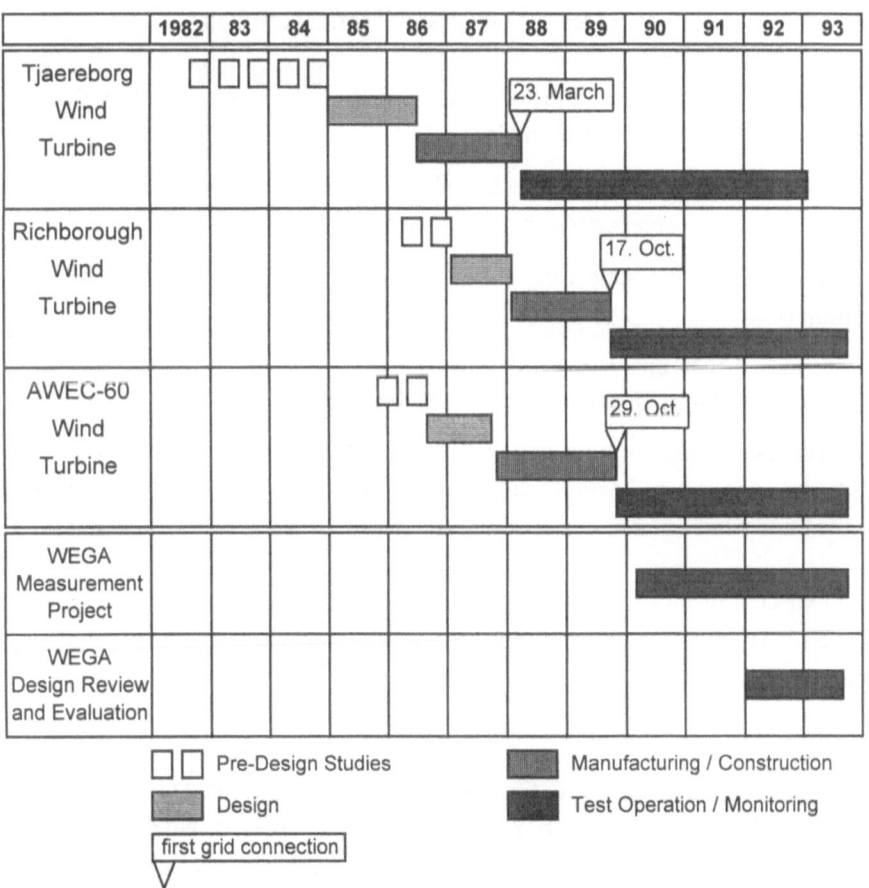

Fig.2.18 Time Schedule of the WEGA Large Wind Turbine Programme

2.4   Data Acquisition and Measurement Programme

The contractors of the wind turbine development programmes have been required to install a scientific oriented data acquisition system and to perform a measurement programme (Fig.2.19). Data processing and evaluation has been coordinated within the framework of the **"WEGA Measurement Project"** managed by ELSAMPROJEKT A/S (Denmark).

The purpose of the programme is to organize, carry out and analyse the results from the measurement programmes of the three wind turbines in such a way, that the maximum of information can be obtained through coordinated action between the WEGA projects. At the same time an uniform database is created with the aim of facilitating later research, carried out by future researchers. The standard measurement period is one year. Time for repair, modifications etc. will be added to the time, which then represents one year of "normal operation".

The short-term objectives of the project. i.e. those to be achieved within the time frame of the project, are the following:

- To obtain better and unified technical reports from the three WEGA-projects thus facilitating comparisons between the projects.

- To ensure that the same weighting is given to the essential issues in all three projects.

- To ensure that interesting and relevant questions arising from one of the projects are taken up and investigated in the other projects.

- To create a comprehensive and well-documented collection of time series data describing the aerodynamics as well as the mechanical dynamic behaviour of the three wind turbines.

The expected result of the project is a better understanding of the technical and economic issues involved in the design, operation and maintenance of large wind energy converters.

The long-term objective, i.e. beyond the time frame of the actual project, is that the standardised data base with time series data, which is created as a result of the project, shall be available to researchers in the area of wind energy and aerodynamics and in this way will help to introduce wind energy into the European electricity supply system in an economic and safe way.

Rotor:	sensors for (numbers): • bending moments on blade roots (6) • stresses on the hub and blade flanges (36) • forces in the pitch control mechanism • pitch angle • rotor position
Drive Train and Generator:	• rotor shaft torque • bending moment in the rotor shaft • rpm of high speed shaft • nacelle position • active and reactive power • current, voltage, frequency • auxiliary power
Control System and Operation:	• active power • pitch angle (see rotor) • wind speed • wind direction • rotor shaft speed • gearbox vibration • ice detector on the nacelle counters for: • hours of operation • idling hours • hours released for operation • hours for failure time • starts and stops • grid connected hours • yaw sequences
Tower:	bending moments at top and foot (4)
Site:	2 meteorological towers (90m height) • wind speed sensors (5) • wind direction (3) • air pressure (1) • air temperature (2)

Fig.2.19 Instrumentation of the WEGA Wind Turbines and the Site as Required by the Scientific Measurement Programme.

All data delivered from the WEGA contractors to the commission are delivered in a common file format. The reports prepared by the WEGA contractors under this contract are prepared according to a common **template report** for each subject (Fig.2.20).

ELSAMPROJEKT A/S POWER STATION ENGINEERING Kraftværksvej 53 DK-7000 Fredericia Phone: +45 75 56 44 11 Fax..: +45 75 56 44 77	WEGA PROJECT

Report description

FATIGUE LOAD CHARACTERISTICS

Reference: EP-525 (W500)	Report no: EP91/345	Signature/date: LEJ/cbe	Approved/date: PC 30/5-91 𝒫𝒞

Distribution: LEJ, MAK, PC, PF : EP KSH : AFM/DTH	Keywords: Vindkraft, måleprogram

Summary:

    This paper is a draft report template prepared under EEC contract JOUR-0025-DK (MB).

    The paper describes the expected content of the report on Duty Cycle Measurements.

    In general the report should fulfil the demands outlined in the Recommended Practices for Wind Turbine Testing and evaluation, part 3: Fatigue Loads [1].

    Aspects which are not mentioned in this paper should therefore be reported according to the Recommended Practices.

Fig.2.20 Example of a Template Report in the WEGA Measurement Project.

The data evaluation concentrates on the following areas:

- Power curve sensitivity analysis, based on variations in wind direction, turbulence intensity, yaw error and precipitation.

- Wind energy potential and power production as a function of wind direction.

- Statistics relating to operation and maintenance costs.

- Structural response and design load verification, as a function of wind speed, wind direction and gustiness.

- Structural response on control activities e.g. power regulation.

- Aerodynamic response of rotor to pitch angle changes.

The evaluation of these data forms the basis for the WEGA Design Review and Evaluation (see chapter 2.5).

Association of GAMMA-60 Evaluation

The Italian GAMMA-60 wind turbine, built by AERITALIA, is in the same class as the three present WEGA projects (see chapter 1). However, this turbine represents a different technical concept (2 blade teetering rotor) and a comparison of experience and measured data with the WEGA turbines is of particular technical and scientific interest.

For this purpose a contract has been arranged by the Commission to allow AERITALIA to carry out work in association with the WEGA Measurement Project. The financial support of the Commission is 400,000ECU. The measurement programme at the GAMMA 60 should follow the guidelines of the WEGA data acquisition system. On this basis an exchange of information and data is being carried out.

The GAMMA 60 wind turbine was commissioned in May, 1991.

2.5   WEGA Design Review and Evaluation

The measured data needs to be compared with the design criteria before the conclusions for an improved design can be drawn or a next generation of large wind turbines can be envisaged. With this aim in view the WEGA Design Review and Evaluation Programme has been initiated. This is a logical extension of the Measurement Project and is coordinated by ELSAMPROJEKT. The following strategy has been established in a close cooperation with the designers of the turbines and associated engineering organisations.

- **Load measurements:**

  Loads are measured under the WEGA-measurement programme and reported according to common report templates.

- **Comparison with design criteria:**

  The load cases, the essential design loads and the material properties are compiled for each turbine. They will be compared with the measured loads.

- **Identification of safety margins:**

  For components and sub-systems included in the design review, safety margins used in the mechanical design will be identified together with the influence of the national standards on allowable stress levels and presumed lifetime of the components.

- **Recommendations for design improvements and cost reductions:**

  On the basis of a comparison between measured loads and design loads, and the ideas about an improved component design the cost reduction potential of the wind turbines through component and sub-system design improvements will be summarised. The improved design shall be of the same basic concept as the turbines. Each party will present recommendations for an improved design of components and sub-systems. The recommendations has to be based on the precondition that the requirements of the national standards are fulfilled.

The design review will comprise the following components and sub-systems:

-   Rotor blades
-   Blade bearings
-   Hub
-   Pitch mechanism
-   Main shaft, hub flange and bearings
-   Gearbox
-   Rotor brake
-   Electrical power system
-   Bedplate and yaw-bearing
-   Yaw drive system
-   Nacelle housing
-   Tower

The WEGA design review and evaluation is still an ongoing programme. The results will be published in a special report in 1993.

Association of the LS-1 Wind Turbine

All three WEGA turbines are of the same basic concept and there is a considerable interest in drawing a comparisons between them and a wind turbine of similar size but of a different technical concept such as the AERITALIA GAMMA 60 and the LS-1 operated by Scottish Hydro-Electric. The LS-1 is a two-bladed turbine with a teetered hub and narrow band variable speed control. It has been built by Wind Energy Group (WEG) and operated on the Orkney Islands during the same time period as the WEGA machines (see chapter 2.1). Comprehensive data acquisition and evaluation has been performed with support of the United Kingdom British Department of Energy.

For the comparison within the WEGA Design Review and Evaluation information on the following subjects have been made available:

-   Load cases
-   Design loads and measured loads
-   Safety margins
-   General design improvement ideas

3
The WEGA Wind Turbines
- Design and Construction

The description of the design and construction of the WEGA wind turbines is based mainly on the 'As-built' and 'Final Design Reports' of the manufacturers or operators of the machines. For this reason the specific reports are not indicated specifically when they are quoted but they are listed in the references.

In order to make the descriptions as comparable as possible a common definition of sub-systems has been applied. According to a widely used definition the subsystems appear as following:

- Rotor blades

- Mechanical drive train consisting of rotor hub, blade pitch mechanism, rotor shaft and bearings, gearbox and high speed shaft to the electric generator

- Nacelle including of the load carrying bedplate, the yaw system and the cladding

- Electrical system consisting of the electric generator, the power electrics for frequency conversion of the variable speed machines and the control and monitoring system

- Tower and foundation

3.1. Tjaereborg 2MW Wind Turbine

The 2MW Wind Turbine at Tjaereborg near Esbjerg, Denmark, was the first large wind turbine to be included in the WEGA large wind turbine programme. It had already been erected in 1987.

3.1.1 General description

The 2MW wind turbine is a 3-bladed, horizontal axis machine, with the rotor placed upwind (Fig.3.1.1). The nominal rotor diameter and the hub height are both 60m.

The blades are cantilevered and suspended at the base of the blade in a plain bearing. The power is controlled by pitching the blades. The blade pitch control system and the yaw system are hydraulic. The hub is a flanged pipe-joint type, with its parts welded together. The equipment for the pitch control system is placed in the hub and supplied by means of a rotating oil union and a rotating shaft. The rotating oil union is specially designed and manufactured for this wind turbine. The hydraulic power supply for yaw and blade pitch systems is placed in the nacelle and has been designed as a complete unit, whose parts can easily be replaced.

The rotor shaft is a long bedplate mounted arrangement with independent front and rear bearings. The gearbox is an epicyclical unit. The first two stages are epicyclical and the third stage is parallel.

The electrical system consists of an induction type generator with an enlarged electrical slip.

The nacelle has a box-like design. It is constructed around a steel-frame fastened to the base frame. The outer coating is galvanised plate. The nacelle fairing is optimised for noise damping.

The tower is made of concrete. The general appearance has been designed in cooperation with an architect to give the large structure a reasonably light look.

The wind turbine is located in Tjaereborg near Esbjerg at the west coast of Jütland in Denmark.

Fig.3.1.1 Tjaereborg 2MW Wind Turbine

Rotor
Number of Blades............................ 3
Orientation............................ upwind
Diameter................................. 61.1 m
Hub Height .... 61 m
Tilt angle.....................................3°

Performance
Rated Rotor Speed.............22.36 rpm.
Wind Speeds: Cut in................5 m/s
                     Rated ....15 m/s
                     Cut out............25 m/s
Design Tip Speed Ratio................. 7.3
$c_{p,R\,max}$.................................... 0.485
$c_{p,R\,max}$ occurs at ....9.8 m/s
Power Control.............. variable pitch
                                   full span

Rotor Blades
Material .... GFRP
Airfoil Section ......NACA 4412-4443
Cord Tip/Root................0.9 m / 3.3 m

Hub
Type..rigid
Material .... welded steel

Gearbox
Type...............2*epicyclic, 1*parallel
Ratio 1:68.4

Shaft Brake
Type.................. disc, parking brakes
Position................... high speed shaft
Operation ....hydraulic

Generator
Type..............induction, enlarged slip
Rated Power ....2,000 kW
Rated Voltage .... 6,000 V
Nominal speed..................1,500 rpm.

Yawing System
Operation ....hydraulic
Yaw Rate ....0.4 °/s

Nacelle
Base Frame .... welded steel
Fairing ....steel frame and plates
Spinner..................... GFRP/sandwich

Tower
Material .... reinforced concrete
Diameter (Top/Base).........4.3m/7.3m

Masses
Blade (each)................................ 9.0 t
Towerhead.............................. 224.6 t
Tower...................................... 665.0 t
Total....................................... 889.6 t

Fig.3.1.2: Main Data of the Tjaereborg 2MW Wind Turbine

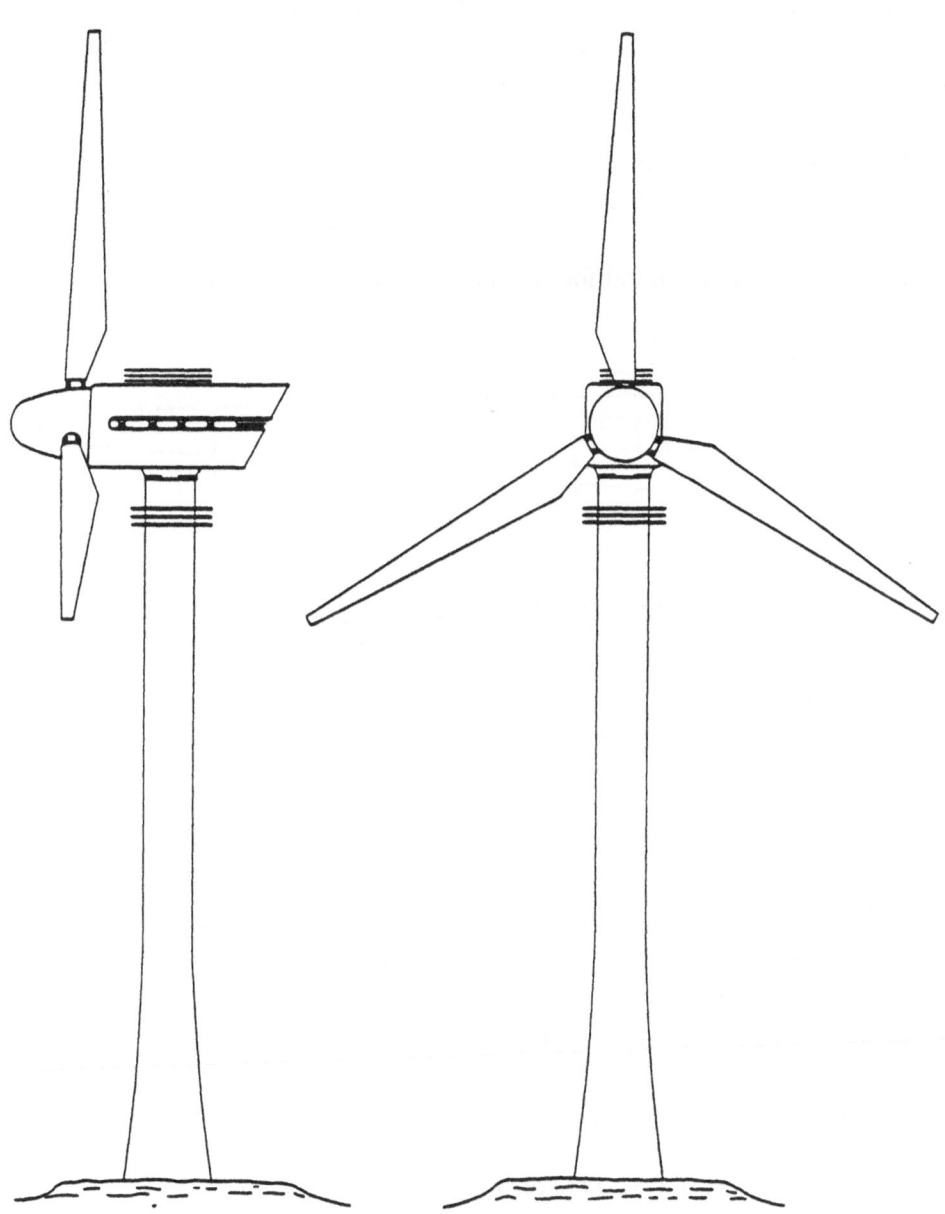

Fig.3.1.3 Tjaereborg Wind Turbine Side and Front View 1:500

3.1.2 Rotor Blades

The blade geometry is designed for a tip speed ratio of 7.3 which is reached at a wind speed of 9.8m/s. The linear tapered projection with a length of 29m has a chord variation from 3.3m at a distance of 5m radius from the flange to 0.9m at the tip.

The twist ratio is 1 degree/3m (from 6° to -3°) and the relative thickness varies between 43% to 12% with a hyperbolic distribution.

The selected aerodynamic airfoil is from the NACA 44XX family.

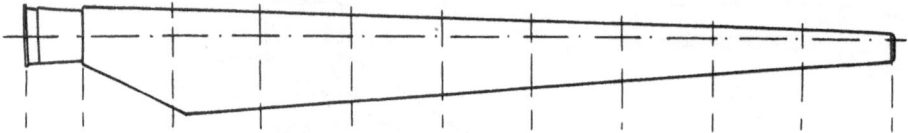

Radius, m	7.0	10.0	13.0	16.0	19.0	22.0	25.0	28.0	31.0
Chord Lenght, m	3.30	3.00	2.70	2.40	2.10	1.80	1.50	1.20	0.90
Thickness, m	1.01	0.72	0.57	0.45	0.35	0.28	0.22	0.16	0.11
Relative Thickness	31%	24%	21%	19%	17%	15%	14%	13%	12%
Twist from Tip, °	8.0	7.0	6.0	5.0	4.0	3.0	2.0	1.0	0.0

Fig. 3.1.4 Rotor Blade Geometry (Dimensions in mm unless otherwise stated)

Each of the rotor blades consists of a forged steel attachment onto which prefabricated shells are glued. The cantilever spar of the roter blade is then wound onto the outside of the shells. The aerodynamic shape is obtained by gluing a curved and a flat shell onto the spar.(Fig.3.1.5)

The rotor shells are made like a sandwich construction: Polyester reinforced with glass fibre on the outside, then a core material of either foam or balsa, and on the inside another layer of glass fibre-reinforced polyester.

The steel attachment is made of forged steel, it is conical and flanged against the root to be fastened to rotor bearings/hub. The weight of the steel flange is 2,600t.

The bearings are of a pre-stressed ball bearing type with an outer diameter of 1,936mm.

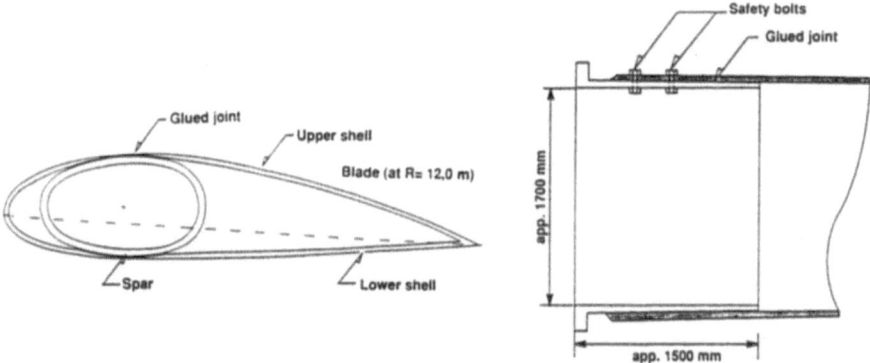

Fig.3.1.5 Cross Section of the Rotor Blade and Steel Flange Attachment

Fig.3.1.6 Manufacturing of a Rotor Blade

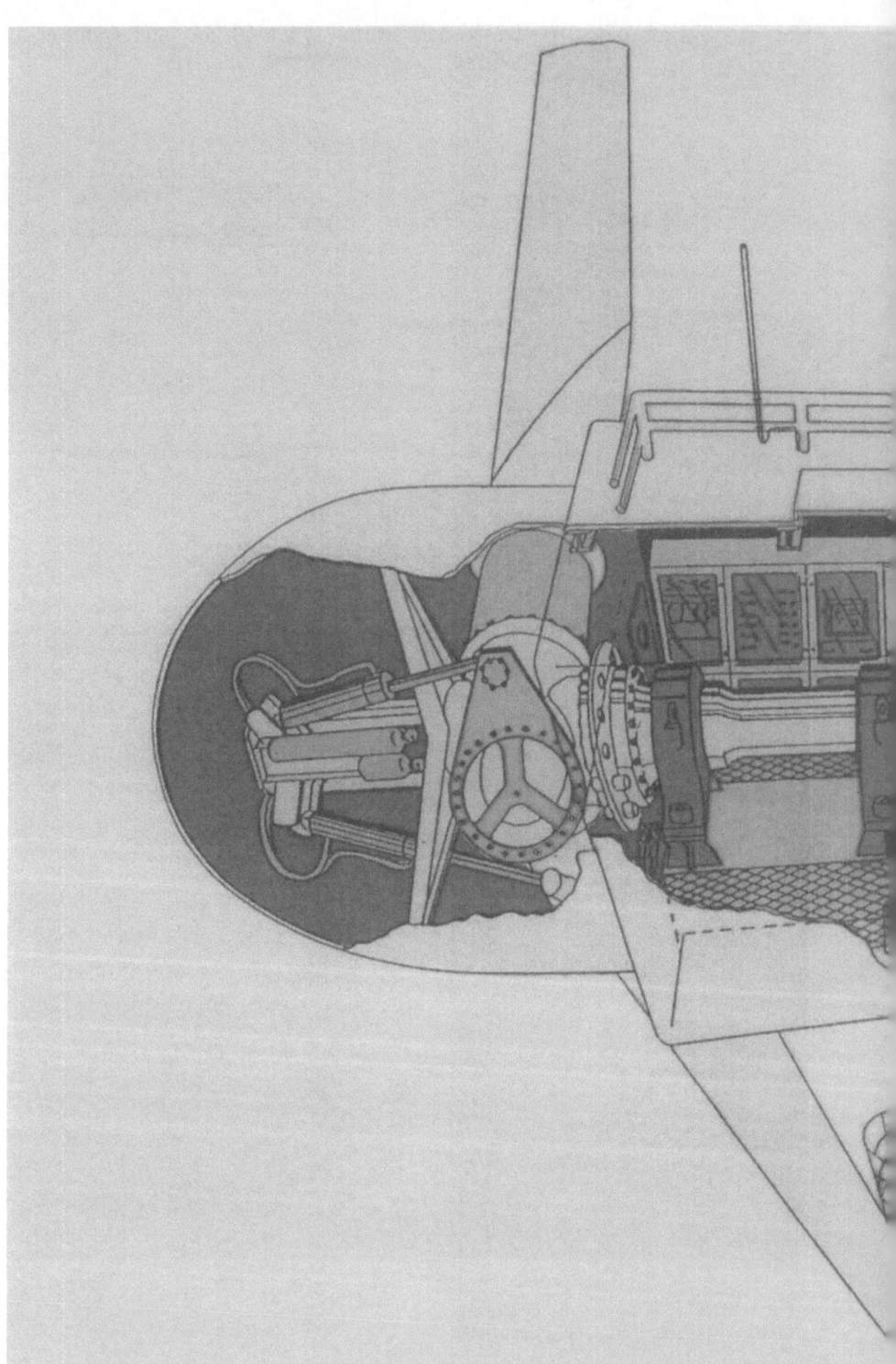

Fig.3.1.7 Artists View at the Nacelle of the Tjaereborg 2MW Wind Turbine

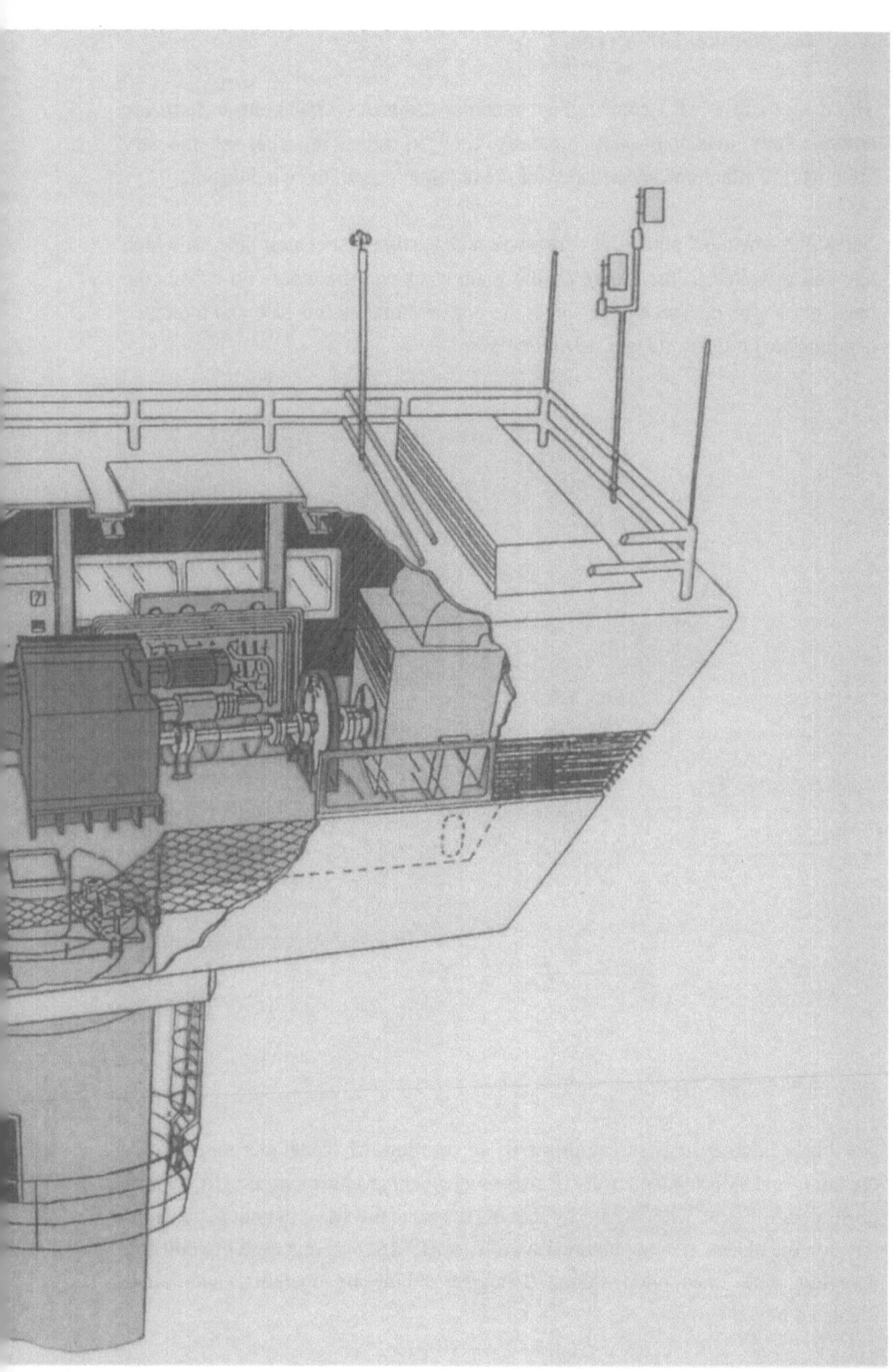

3.1.3 Mechanical Drive Train

The drive train is of a conventional modular design. All components used are commercially available ones normally used in other branches of industry although manufactured specifically for the requirements of the wind turbine.

The hub is a welded pipe-joint construction consisting of a central pipe, of which one end is bolted to the flange on the main shaft and the other end carries the blade pitch system. Three larger pipes are placed radial on this pipe and the blade bearings are bolted to flanges on these pipes.

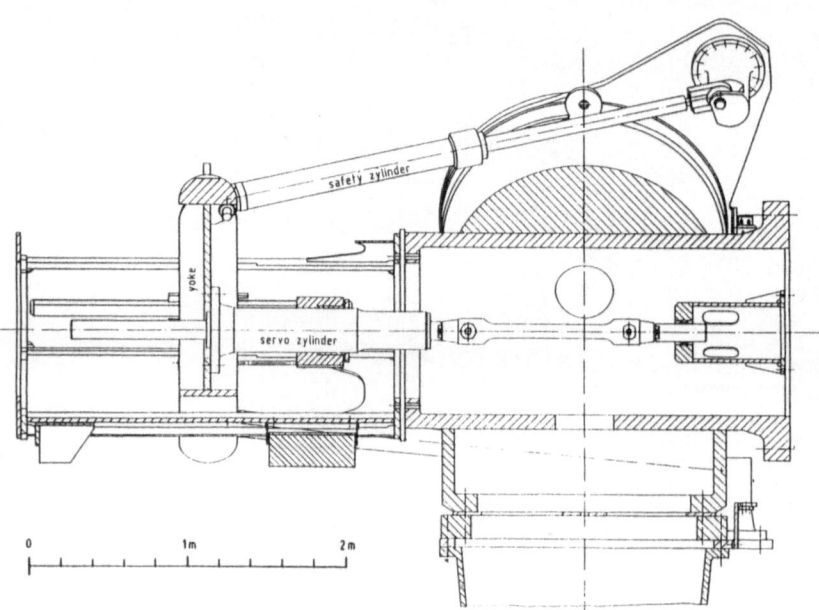

Fig.3.1.8 Blade Pitch System

The blade pitch system serves primarily to position the blades during start-up, operation and stop sequences. It consists of a tandem arrangement of cylinders, as illustrated in figure 3.1.8, where by moving the yoke the servo cylinder is used to position the blades in the operating range -2 to +35 degrees. It is used both during operation with fixed pitch around 0 degree and during operation with pitch regulated power control.

The oil flow to the servo cylinder comes from a electrically driven pump and it is controlled by a servo valve which receives its control signal (oil rate demand) from the control computer.

Fig.3.1.9 Hub and Blade Pitch Mechanism

The forged hollow main shaft is supported by two spherical roller bearings as indicated in figure 3.1.11. The front bearing is a radial one whereas the rear one is of the axial/radial type which takes up the thrust from the rotor.

The gearbox is a combined epicyclical/parallel one with two epicyclical stages and one parallel final stage with the output shaft shifted from the centreline of the epicyclical stages. This design gives through the gearbox and the hollow shaft access to the hub for cables and hydraulic pipes. It is further equipped with an inching device, for service and maintenance, consisting of an electric motor with a spring activated electromagnetic lifted brake and a manual clutch.

The high speed shaft (1,500rpm) carries the parking brake and transmits the torque from the gearbox to the asynchronous generator. The generator has a special high slip design, with a high resistance rotor, in order to give sufficient damping towards torsion oscillations in the power train.

Fig.3.1.10 Mechanical Drive Train during Assembly

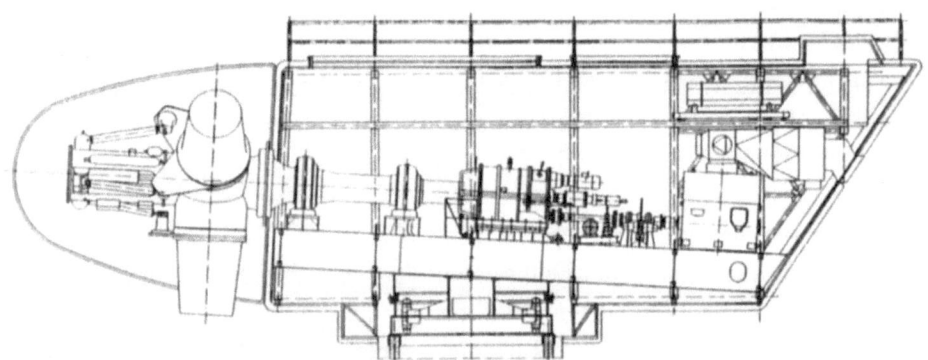

Fig.3.1.11 Nacelle and Drive Train Arrangement

3.1.4 Nacelle Structure and Yaw System

The main supporting structure in the nacelle is a welded two-cell box girder called the base frame. Height, width and length of the base frame are 1,000mm, 1,900mm and roughly 15,000mm, respectively. Over the tower, the base frame is of circular shape with the same diameter as the tower.

The topside of the base frame forms the bed plate of the main machinery, such as main bearings, gear and generator.

The nacelle is constructed around a steel frame fastened to the base frame. The outer coating are galvanised plates. A noise absorbing plate is placed on the inside, and a layer of mineral wool is placed in between. The nacelle is connected to the support structure only via vibration dampening mounts to prevent structural noise from the nacelle .

At the bottom of the nacelle an emergency exit leads to a ladder on the outside of the tower. This ladder goes down to a balcony. From there, a door opens to an internal ladder in the tower. The top of the nacelle is equipped with a duckrun from where it is possible to inspect the central parts of the blades and the upper part of the nacelle coating.

The hub and the blade pitch mechanism are covered by a spinner. The spinner is a sandwich construction with an internal and an external laminate with a light core material.

The yaw system serves to turn the nacelle into the wind and to keep the rotor plane closely aligned perpendicular to the average wind direction at any time. The system is hydraulically operated and the nacelle is turned in the wind by two hydraulic motors located in wells under the floor of the nacelle. The motors are equipped with three stage epicyclical gearboxes and cogwheels and they drive on pinions on the outer rim of the yaw bearing. The drive motors are on-off controlled by the computer and there are further six hydraulic brake callipers to keep the nacelle firm in its position. The system is powered by a separate pump.

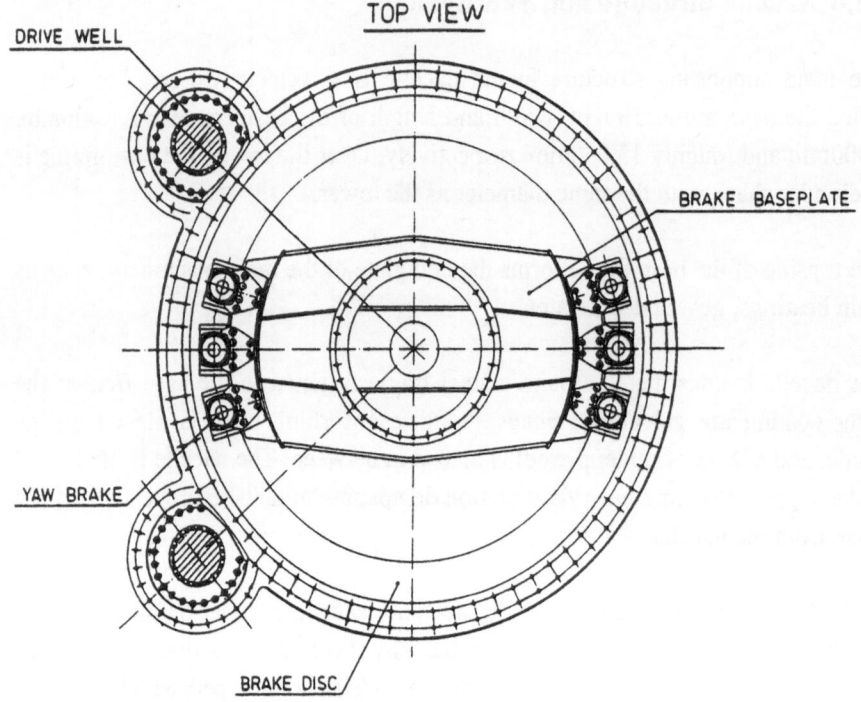

Fig.3.1.12 Components of the Yawing System

3.1.5 Electrical System and Control

The **Generator** is a 4-poled induction generator with a squirrel cage rotor chosen because of its simplicity and low maintenance requirements. To reduce the dynamic loads a generator with an enlarged slip of 2%, realised with a specially designed resistance rotor, at the cost of approximately 1% lower efficiency has been chosen.

Rated voltage.. 6.0 kV
Rated power .... 2,000 kW
Rated current.. 221 A
Rated power factor.. 0.87
Rated speed .... 1,521 rpm
Rated efficiency .... 96 %

Fig.3.1.13 Main Data of the Generator

A one line diagram of the electrical power system is shown in figure 3.1.14. The diagram is kept as simple as possible. As circuit breakers are used magnetically held SF6 contactors in order to make the cut out of the generator as reliable as possible. The turbine is equipped with a 350kVar capacitor bank to compensate for the reactive power required at zero load.

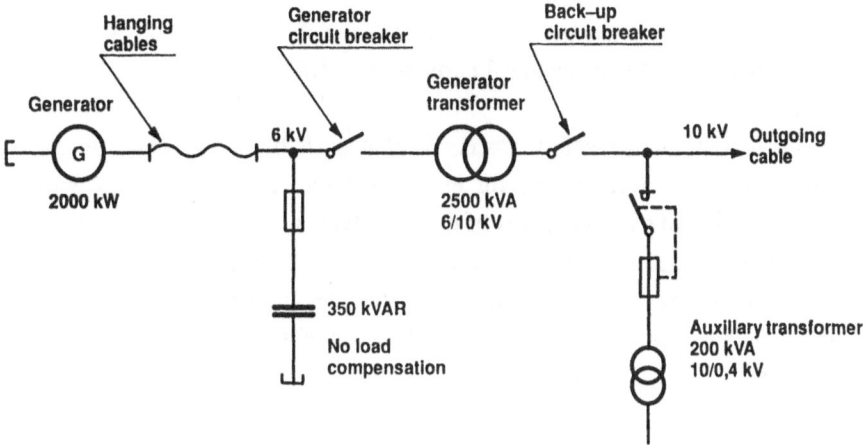

Fig. 3.1.14: Electrical System

The **Control System** is structured into 4 main functional blocks.

The first block is the sequence control program for the rotor system. The program starts and stops the turbine and chooses between either fixed or pitch operation at low wind speeds. It also controls the power by pitch regulations at wind speeds over the rated wind speed. The primary transducers used for this purpose are two pitch angle transducers in the hub and the two power transducers in the tower.

The second block comprises the sequence control of the yaw system. The primary transducers for this task are the two wind direction transmitters mounted on top of the nacelle.

The third block interfaces the input from the analog transducers to the two sequence blocks. It scans and scales the signals, performs the various averaging functions and generates the status information (limit values) and alarms used by the two sequence blocks.

The last block performs all "continuous" control functions for the rotor system, for instance pitch regulation (fixed pitch, pitch rate) and power control. Switching between the functions is controlled by the sequence block.

The design philosophy of the **Safety System** is:

-   The safety system shall alone be able to protect the turbine under all circumstances and to bring it to a safe idling speed.
-   The control system supervises the transducers during operation and tests the function of the mechanical parts of the safety system.

The basic feature of the safety system is that each blade has its own safety cylinder with a gas accumulator for energy storage. The safety cylinders are relieved through two parallel valves of different type and each cylinder can bring its associated blade to a position between 55° and 90°. The system is tolerant to one fault as all transducers and circuits are redundant and since two blades in the 55° to 90° range, even if one should be stuck, are enough to bring the turbine to a safe idling speed.

There is further a mechanical disk brake which can be activated manually in case of emergency. It can stop the turbine without aerodynamic braking.

3.1.6 Tower and Foundation

The tower of the Tjaereborg Wind Turbine is made of non prestressed but steel reinforced concrete.

Material	reinforced concrete
Height of power	57 m
Wall thickness	250 mm
Outside diameter, root	7.41 m
Outside diameter, centre	4.75 m
Outside diameter, tower neck	4.25 m
1st natural frequency	0.85 Hz
2nd natural frequency	5.44 Hz
Access	internal lift or ladder
Foundation	hexagonal plate found

Fig.3.1.15 Main Data of the Tower

At the base, the tower geometry is a parabola gradually turning conical, its last approx. two meters end in a cylinder in which the rods transfer the loads to the tower through the tower neck.

The nacelle can be reached through a rack-and-pinion lift. Furthermore, there are emergency ladders throughout the tower. The emergency exit from the nacelle consists of an external ladder going down to the external gallery which leads to the internal ladder.

The tower interior has four bottom levels, from the top: Measuring equipment room, relay equipment room, HV room, and at the bottom the transformer room. Furthermore, there are 4 internal steel platforms located at the same height as down-turned blade tip, centre of blade, highest lift stop, and finally, the rotating stairs leading to the nacelle.

The tower is directly founded on a hexagon reinforced concrete plate whose width between side lines is 22m and height is 2m.

3.1.7 Assembly and Erection

The design of the turbine was started in 1985 and the main components were procured so that shop assembly could start in September 1986 when the bed frame was delivered from the manufacturer to the power station Vestkraft. Here it was placed in a ground level passage under the generator building where the station service crane could have access through hatches to the upper floor.

At this place the heavy main components (low speed shaft with main bearings, gearbox, high speed shaft with brake disk and bearings and the generator) were mounted onto the bed frame utilising the service crane for the heavy lifts. Afterwards the entire bed frame was placed on a flat car by the service crane and moved to the site where it was placed onto a steel rack in a provisional assembly hall.

Here the assembly was completed. The remaining components and auxiliary systems were mounted on the bed frame and the hub and blade pitch system were bolted onto the main shaft flange. In this phase all electrical systems as well as the control system were installed. Finally the nacelle cladding and the spinner were mounted on the bed frame.

Prior to lifting, a thorough checkout and functional test of all systems was carried out and the drive train was tested at rated speed by an electrical variable speed motor driving by means of a belt on one of the couplings on the high speed shaft. During this test the blade pitch system was in action, monitored by the control system under a special test programme. The nacelle was ready for lifting early in September 1987.

Parallel to assembly and testing of the nacelle, the tower was constructed as a slip form poured concrete structure and the installations, mainly electrical systems and a lift in the tower were completed as far as possible.

Early in the project a lot of attention was drawn to the problem of placing the nacelle on the top of the tower and a number of possibilities were taken into consideration, regarding both technical feasibility as well as costs. Two possibilities remained at last and it was finally decided to use 4 steel towers and a hydraulic jacking system, lifting two beams under which the nacelle was hanging.

Fig.3.1.16 Assembly of the Nacelle

When the nacelle was lifted as high as to be passed over the tower, it was pushed into place with hydraulic jacks and lowered onto the tower. This method permitted lifting in higher winds than a mobile crane could, though at a somewhat higher initial cost. If high winds had prevailed, with a mobile crane at the site, its high rates would have made this method the more expensive one.

Lifting commenced on Thursday September 18, 1987 and lasted 14 hours with a midway stop overnight. The wind speed was about 10m/s during most of the lift, increasing to 15m/s at the end of the operation.

The blades were mounted by two mobile cranes on November 6, 1987.

Fig.3.1.17 Erection of the Turbine

3.2   Richborough 1MW Wind Turbine

The Richborough Wind Turbine, developed and manufactured by James Howden & Company Ltd., is based on the successful development of a range of machines with ratings between 60kW and 750kW and incorporates similar design principles, particularly a rotor constructed of wood/epoxy laminates, which is light in weight and enables the tower head weight to be kept to a minimum (see chapter 2).

3.2.1 General Description

The wind turbine is a three blade, pitch tip controlled machine with a rated power output of 1,000kW at a wind speed of approximately 12m/s. The three blade upwind rotor is used for a number of reasons, for instance smooth power output, good start up characteristics, and reduced dynamic blade loads.

The rotor diameter is 55m and the hub height of the machine 45m.

The free standing tubular styled tower, which is fabricated in steel, flares towards the base in order to increase the foundation bolt pitch circle diameter, and to provide accommodation for all the necessary electrical and control equipment.

The basic design concept incorporated a number of innovative features. Its essential items are:

-    lightweight wood/epoxy blades

-    thick airfoil sections near the blade roots, to allow weight savings to be made

-    hydraulicaly activated movable blade tips

As with the other WEGA wind turbines a modular drive train arrangement with two main rotor bearings and an epicyclical gearbox was chosen. The main rotor speed is fixed, because the 6-pole induction generator is directly grid connected.

The machine is located at the Powergen's Richborough Power Station in Kent. The site is 2km from the cost and is mostly level.

Fig.3.2.1 Richborough 1MW Wind Turbine

Rotor

Number of Blades............................. 3
Orientation............................. upwind
Diameter .... 55 m
Hub Height .... 45 m
Tilt angle..0°

Performance

Rated Rotor Speed................24.5 rpm
Wind Speeds: Cut in................5 m/s
               Rated ....12 m/s
               Cut out............25 m/s
Design Tip Speed Ratio................. 7.1
$c_{p,R\,max}$.................................... 0.447
$c_{p,R\,max}$ occurs at ....9.9 m/s
Power Control............... variable pitch
                 blade tip (4 m)

Rotor Blades

Material ....wood/epoxy
Airfoil Section ....LS(1)-0421 Mod.
Cord Tip/Root................0.8 m / 3.2 m

Hub

Type......................rigid, bolted flange
Material ....cast steel

Gearbox

Type............... 1*epicyclic, 1*parallel
Ratio 1:40.8

Shaft Brake

Type...............disc, emergency brakes
Position...................... low speed shaft
Operation.....................spring applied,
              hydraulically released

Generator

Type..................................... induction
Rated Power ....1000 kW
Rated Voltage......................... 3300 V
Nominal speed....................1000 rpm.

Yaw System

Operation............................. hydraulic
Yaw Rate ....0.6 °/s

Nacelle

Base Frame........... plate girder (Steel)
Fairing .... GFRP

Tower

Material ....steel
Diameter (Top/Base).....3.0 m / 6.0 m

Masses

Blade (each)................................. 3.6 t
Towerhead................................. 83.4 t
Tower.. 76.5 t
Total.. 159.9 t

Fig.3.2.2 Main Data of the Richborough Wind Turbine

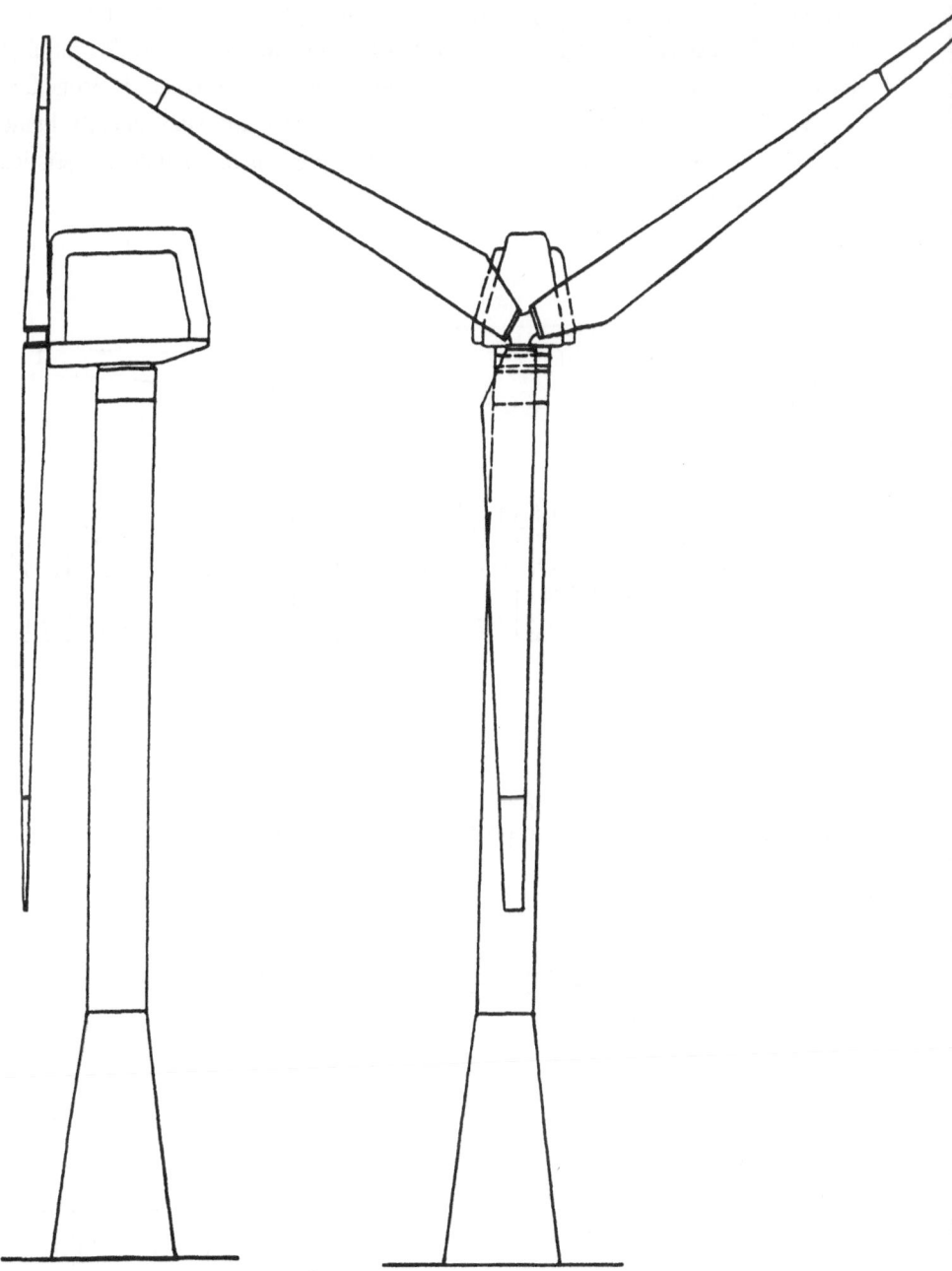

Fig.3.2.3 Side and Front View of the Richborough Wind Turbine 1:300

3.2.2 Rotor Blades

The airfoil section used, LS(1) - 0421 Mod, is a high performance type for low speed applications, with superior performance throughout the Reynolds number range. From mid-span inwards, the trailing edge of the blade is truncated (Fig.3.2.4) to effectively increase the section thickness, and subsequently blade strength. This method of section thickening is favoured, as the stall characteristics of the airfoil are unaffected.

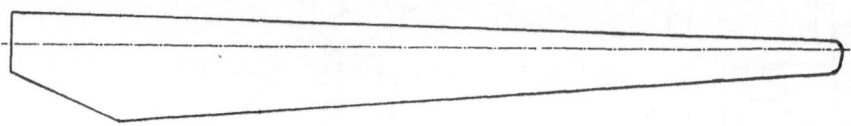

Radius, m	4.3	6.6	8.8	11.0	13.2	15.0	17.6	19.8	22.0	23.5	25.3	27.5
Chord Lenght, m	3.20	2.95	2.71	2.48	2.24	2.05	1.79	1.57	1.35	1.20	1.02	0.80
Thickness, m	1.04	0.90	0.78	0.66	0.53	0.43	0.38	0.33	0.28	0.25	0.21	0.17
Relative Thickness	32%	31%	29%	26%	24%	21%	21%	21%	21%	21%	21%	21%
Twist from Tip, °	13.0	11.7	10.5	9.2	8.0	7.0	5.5	4.3	3.1	2.2	1.2	0.0

Fig.3.2.4 Blade Geometry

The blades are made of wood/epoxy laminates (Fig.3.2.5). The hollow "´D` spar" which follows the aerodynamic profile leads to a light blade with high torsion rigidity.

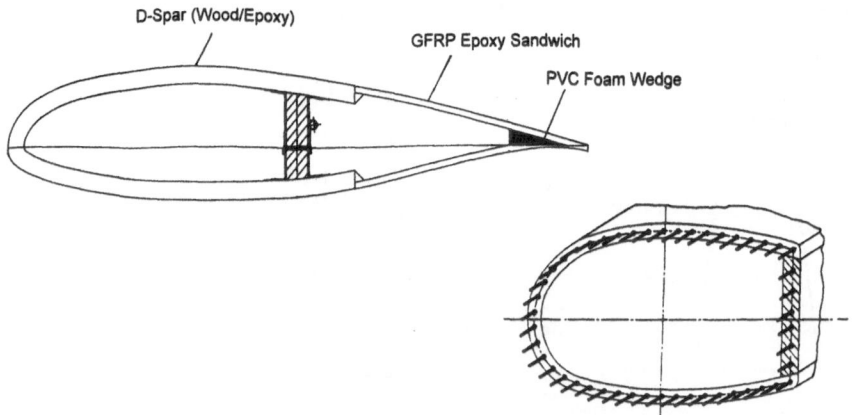

Fig.3.2.5 Rotor Blade Cross Section and Blade Attachment (no Steel Flange)

The main blade root to hub fixing comprises high tensile studs set in a carbon fibre/epoxy grout. No heavy metal flange is needed.

The exposed surfaces of the tip and main blade are covered with woven glass cloth and a modified polyester gel coat, which is resistant to abrasion and UV degradation. The leading edge of the blade is also covered by a 150 mm wide polyurethane anti-abrasion tape.

The blade tip pitch control system is used for stop/starts and is also used for power control, at greater than rated wind speeds.

The blade tip section, which swivels about a compound spar, is mounted radially at the end of the blade and is biased towards the stop position by a pre-loaded torsion bar. A linear actuator mounted at the end of the main blade moves the tip section into the run position against the torque produced by the torsion bar.

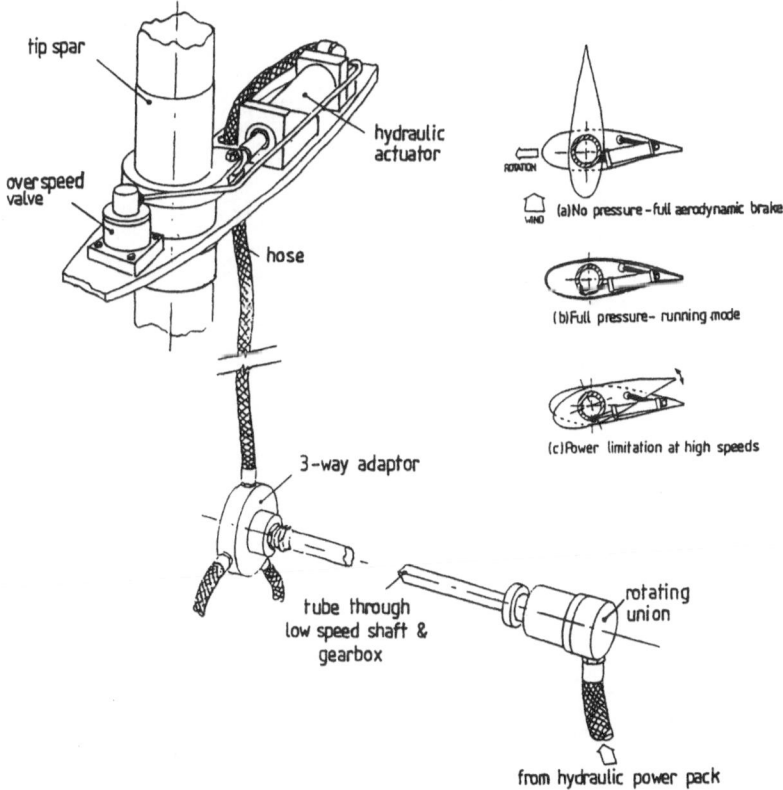

Fig.3.2.6 Tip Pitch Mechanism

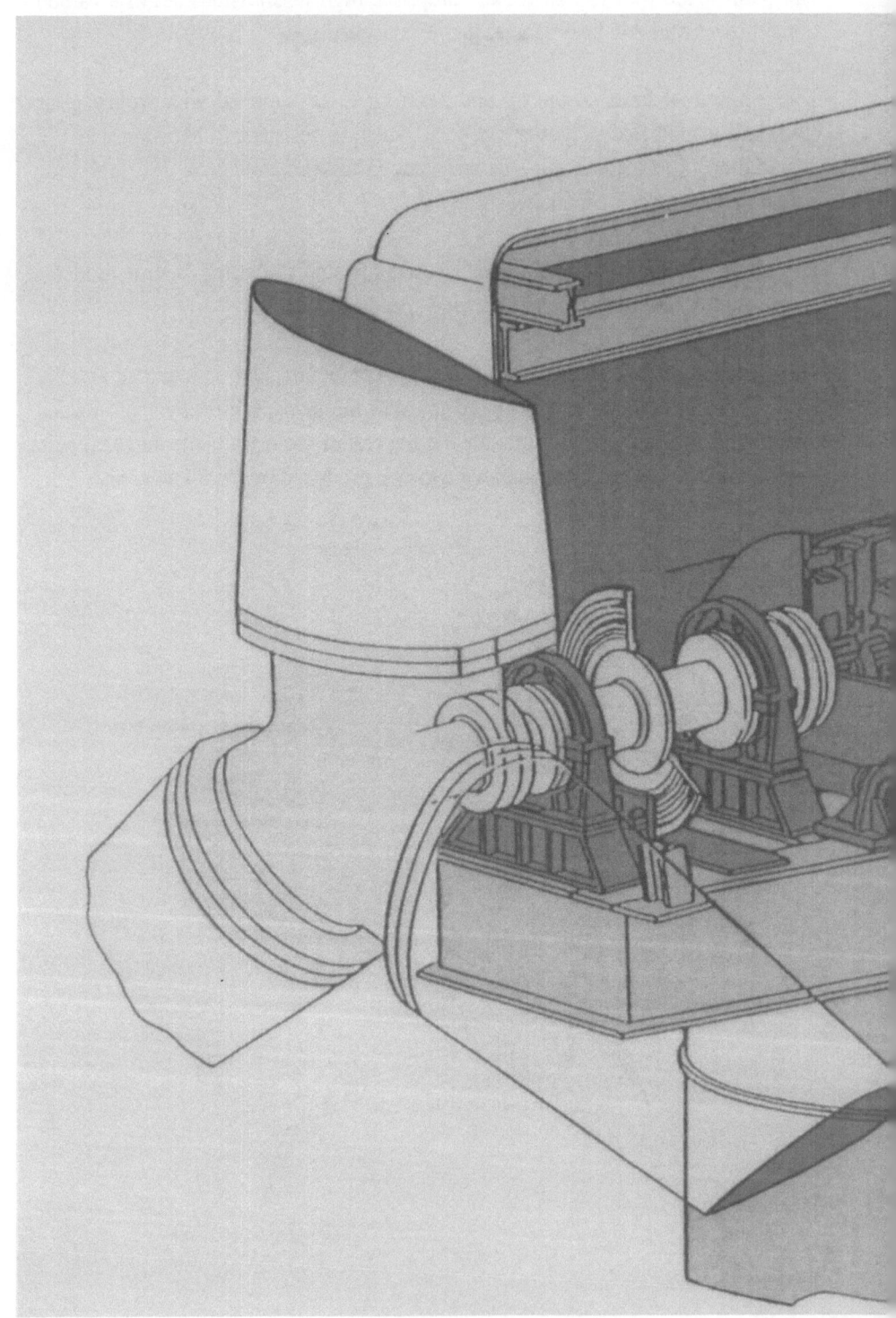

Fig.3.2.7 Artists View at the Nacelle of the Richborough 1MW Wind Turbine

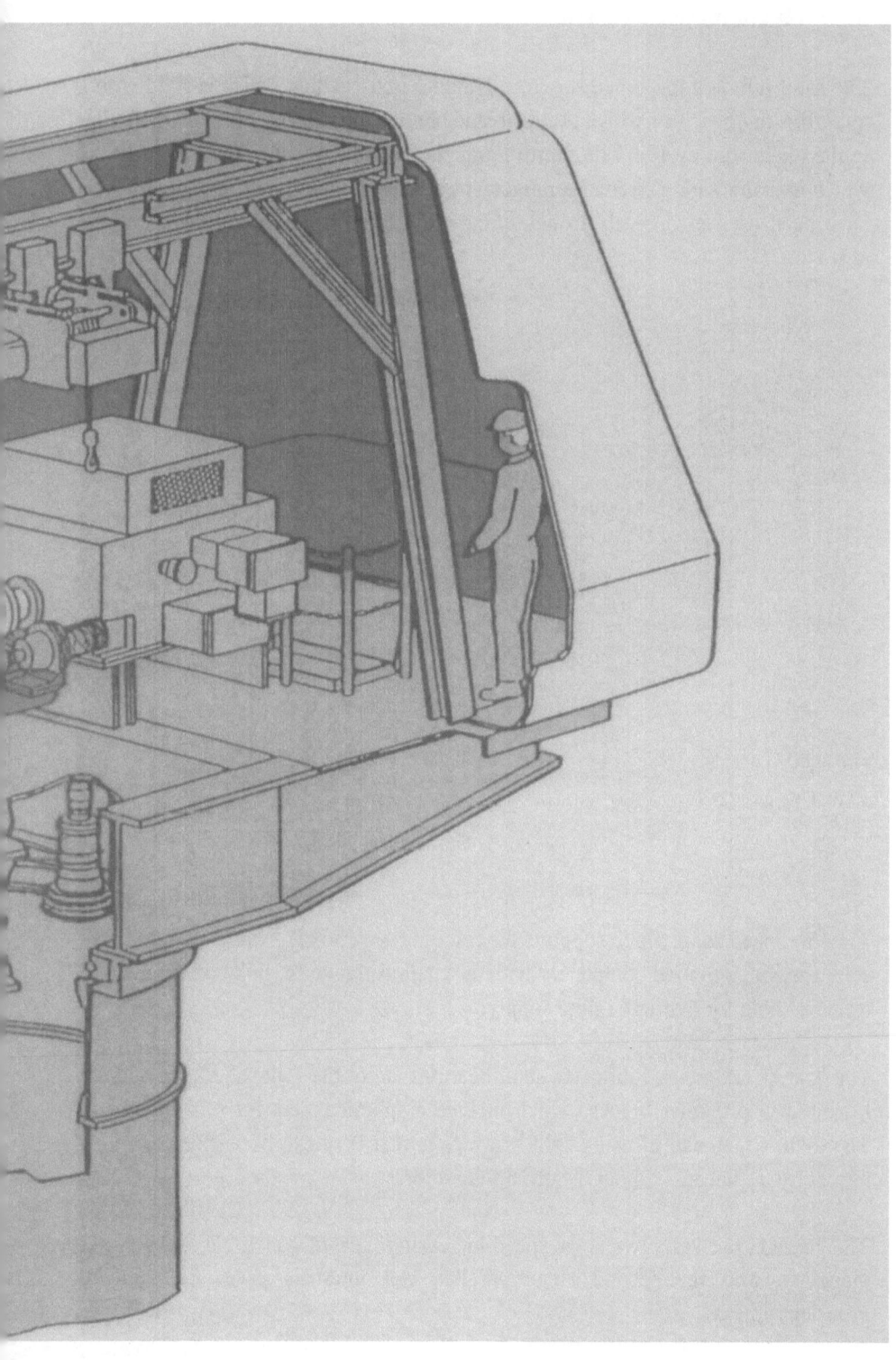

3.2.3 Mechanical Drive Train

The rotor hub is a machined spheroid graphite iron casting, which is rigid with a total of 4 flanges, 3 of which are for blade connection, their profile being similar to the blade root profile. The fourth flange is for the blade hub shaft connection, which is made with a circular arrangement of stud bolts. Finite Element Analysis have been used extensively to determine the fatigue life and deflections of the hub.

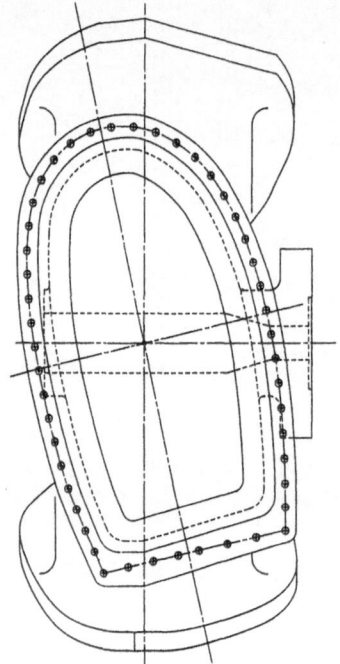

Fig.3.2.8 Hub Flange

The low speed shaft which supports the rotor and gearbox is manufactured from a steel forging, supplied completely with an integral flange for hub connection and a central hole for hydraulic fluid supply.

The low speed, grease lubricated shaft bearings are of the split roller type and are mounted in plummer blocks, which include a spherical seat for self alignment. The main advantage of using split bearings is that they can be inspected, and if necessary, replaced, without disturbing the rotor.

The gearbox used is a two stage speed increasing unit, of which the first stage is planetary, and the second stage parallel. All internals used are standard proprietary items.

A parallel shaft second stage is used to offset the output shaft from the input shaft centre line, which allows a hydraulic supply line for the tip actuators to pass through the gearbox into the low speed shaft. The gearbox is 'splash lubricated', with the circulation oil pump and the air blast cooler being included for cooling purposes only. If either of these components were to fail, the machine would trip on 'over temperature', with no resultant damage to components.

The gearbox input shaft is hollow and is connected to the male end of the low speed shaft by using a standard proprietary shrink disc unit. It also has an integral turning gear for rotor maintenance, consisting of an electric geared motor, meshed with the high speed end of the gearbox via a manually operated clutch. Connection to the generator is made through a spacer type coupling which tolerates reasonable misalignment.

The rotor brake is mounted on the low speed shaft between the bearings. It is a disc brake with spring applied hydraulically released callipers, which can operate at two levels by varying the hydraulic pressure in the system. Normal stops are initiated by the blade tips moving to the stop position, thus reducing the power output of the rotor to zero, and with a reduced braking force. However, when it is necessary to halt the rotor in the minimum time, an emergency stop is performed and the full capacity of the brake is used. To insure complete safety during maintenance periods, a rotor lock device is incorporated, which is manually operated to engage/disengage as appropriate.

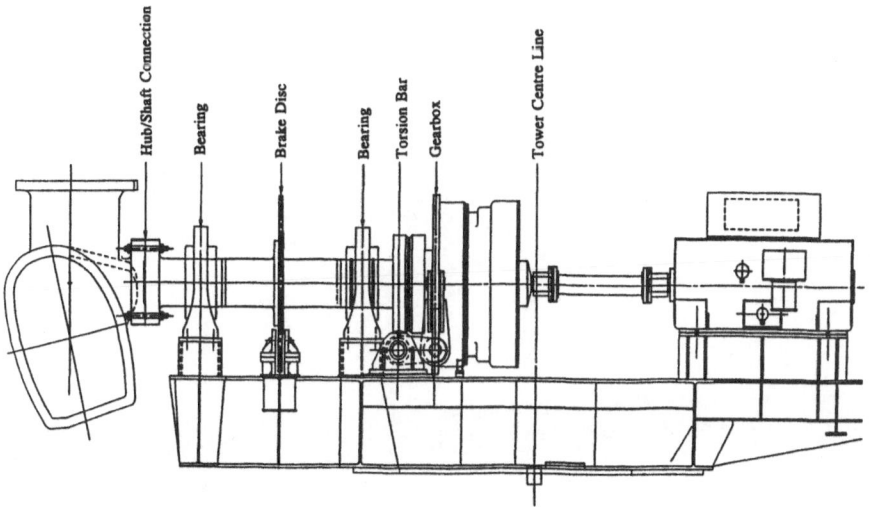

Fig.3.2.9 Mechanical Drive Train Arrangement

All hydraulic components included in the machine are supplied from a single electro-hydraulic power pack, which consists of a reservoir, pump accumulator, and all the necessary valves required for control. The valves and hydraulic components are designed to be fail-safe, with back-up dump valves in the blade tip and rotor brake circuits. In case of a power or pressure loss the machine will close down in a safe manner. The hydraulic power pack is mounted on the machine bedplate.

Fig.3.2.10 Inside View of the Nacelle

3.2.4 Nacelle Structure and Yaw System

The machine bedplate, which supports the drive train and all associated equipment, is fabricated from plate girders. The bottom surface of the plate girders are fixed to a flange suitable for the slew ring. All drive train components are bolted to the top surface via machined pedestals. Cross members are included at points of load application, e.g. bearings. The bedplate is completely made of structural steel.

The nacelle fairing which provides the machine bedplate with protection, is manufactured in GRP and incorporates a number of apertures for access and ventilation purposes. The appearance of the machine is greatly affected by the nacelle form, therefore this aspect has been given careful consideration.

The machine bedplate is connected to the top tower flange via a standard slewing ring. This consists of a large diameter, light section, crossed roller bearing, complete with an internal ring gear, and is grease lubricated.

The yaw drive system comprises a number of hydraulic geared motors mounted local to the slewing ring. A pinion is mounted on the output shaft of each motor, which meshes with the internal gear of the slewing ring. The motors drive the machine bedplate to the correct rotational position relative to the mean wind direction. The yaw brakes keep the bedplate in the correct position to the wind. The brakes are of the callipers type mounted on the underside of the machine bedplate and operate on the inside of the machined top tower flange. They are spring applied, hydraulically released which ensures fail-safe operation.

3.2.5 Electrical System and Control

Three electrical generator options were examined, to determine whether there were advantages to be gained in selecting either a "high efficiency" induction generator, or a two speed pole amplitude modulation induction generator, in preference to a "standard" induction machine.

A six-pole induction generator was selected, with a full load speed of 1,012rpm. The rotor speed is 24.5rpm and the rotor power generated at a rated electrical output is 1,066kW.

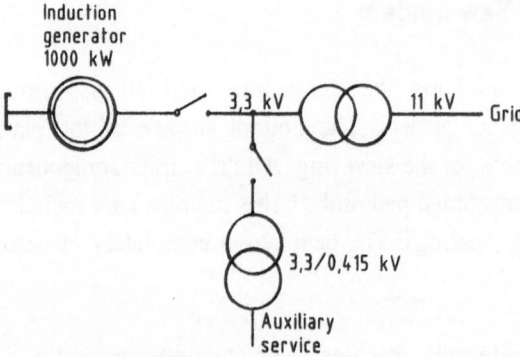

Fig.3.2.11 Electrical System

The control system performs two distinct functions; the monitoring function and the operating function.

The monitoring function is the overall system arbiter, and makes decisions which determine the performance of the operating function. These decisions are based on inputs to the system, and the performance of the system itself. The first decision to be made is whether the system is in the manual or automatic mode. The monitor determines this, based on the position of a front panel control.

Having decided on the mode of operation, the monitor determines the required "state" of operation within that mode. The term "state" is used to describe a particular configuration of the system. The state information which is output from the monitoring function is passed to the operating function as a request. It is then up to the operating function to perform in such a way as to arrive at the required state.

3.2.6 Tower and Foundation

The cylindrical steel tower is assembled from a number of bolted sections and is high, to give a rotor centre line height of 45m. The conical base section simplifies connection to the foundations and permits all the electrical equipment to be housed in the tower base. As in previous Howden designs an access door is provided in the tower, level with the end of the main blade, to enable maintenance to be carried out.

All site joints in the tower are bolted, and the tower is split into sections, each being within normal road transport limitations. All access to the machine bedplate is internal.

Material... steel
Height ....45 m
Diameter, root...6 m
Diameter, cylindrical part ....3 m
Mean Wall thickness ....20 mm
First natural frequency axial (calculated)........................ 1.076 Hz
First natural frequency transverse (calculated)................ 1.095 Hz

Fig.3.2.12 Main Data of the Tower

Fig.3.2.13 Construction of the Reinforced Plate Foundation

3.3   AWEC-60 1.2MW Wind Turbine

The AWEC-60 design is partially based on the WKA-60 wind turbine, which was developed by MAN Technologie, Germany. The two wind turbines share the conception of the nacelle and the mechanical drive train. Extensive differences appear in the rotor blades, electrical generator system, control subsystem and tower.

3.3.1 General Description

AWEC-60 is a three bladed pitch controlled wind turbine. The rotor, with a diameter of 60m, is placed upwind to the tower.The height of the rotor hub is 46m. At a wind speed of approximately 12m/s the rated power of 1,200kW is achieved. The rotor features a full span pitch system for power and speed control. The pitch mechanism operates with hydraulic cylinders.

The rotor blades are made of a fibre glass compound material. The load-carrying spar consists of approximately 80% of unidirectional glass-rovings.

The components of the mechanical drive train are mounted directly onto a supporting bedplate. As far as possible standard components are used i.e. for the gearbox and the rotor shaft bearings.

The electrical power system can be operated speed variable in a speed range of ± 10 % around the nominal rotor speed of 23rpm. This is achieved by an oversynchronous cascade connected to the induction generator. The control system is capable of controlling the generator torque and slip to perform the rotor speed and electrical power control at the speed variable operation.

The nacelle is a steel frame design with a fairing specially designed for damping the mechanical noise of the drive train components.

The free-standing tower is fabricated in steel. A cylindrical upper section leads to a conically shaped base.

The wind turbine is located near Cabo Villano in the Northwest of Spain, directly at the see shore.

Fig.3.3.1 AWEC-60 Wind Turbine

Rotor

Number of Blades............................ 3
Orientation............................. upwind
Diameter..................................... 60 m
Hub Height............................... 46 m
Tilt angle...4°

Performance

Rated Rotor Speed...... 23 rpm ± 10%.
Wind Speeds: Cut in..............5.2 m/s
                        Rated ....12.0 m/s
                        Cut out............24 m/s
Design Tip Speed Ratio................... 7
$c_{p,R\ max}$................................... 0.465
$c_{p,R\ max}$ occurs at ....10 m/s
Power Control............. variable pitch,
                                        full span

Rotor Blades

Material .... GFRP
Airfoil Section ....NACA 44XX-series
Cord Tip/Root............... 1.0 m / 3.5 m

Hub

Type..rigid
Material .... steel, casted

Gearbox

Type................2*epicyclic, 1*parallel
Ratio 1:47.8

Shaft Brake

Type............... disc, emergency brake
Position..................... low speed shaft
Operation .... hydraulic

Generator

Type........................... induction with
                        oversynchronous cascade
Rated Power ....1,200 kW
Rated Voltage .... 6,300 V
Nominal speed...................1,100 rpm.

Yaw System

Operation .... hydraulic
Yaw Rate .... 0.8 °/s

Nacelle

Base Frame .... welded steel
Fairing ....steel frame and plates
Spinner...................steel frame/GFRP

Tower

Material ....steel
Diameter (Top/Base).......3.5 m / 8 m

Masses

Blade (each)............................... 6.6 t
Towerhead............................... 189.6 t
Tower...................................... 92.0 t
Total...................................... 281.6 t

Fig.3.3.2 Main Data of AWEC-60

Fig.3.3.3 Side and Front View of AWEC-60 1:500

3.3.2 Rotor Blades

The tapered rotor blades are 28.1m long, with the chord varying from 3.4m at the inner aerodynamic section (relative thickness 33%) to 1.0m at the tip (relative thickness 12%). A linear variation of the twist angle leads to a total twist of 9°. The selected airfoil is from the NACA 44XX family. The aerodynamic layout refers to a tip speed ratio of 7 at a corresponding wind speed of 10m/s.

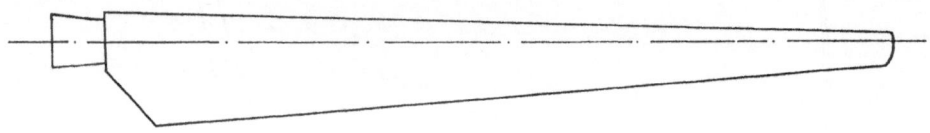

Radius, m	6.0	8.0	10.0	12.0	14.0	16.0	18.0	20.0	22.0	24.0	26.0	28.0	30.0
Chord Lenght, m	3.35	3.16	2.96	2.76	2.57	2.37	2.18	1.98	1.78	1.59	1.39	1.20	1.00
Thickness, m	1.04	0.85	0.71	0.61	0.51	0.45	0.39	0.34	0.29	0.24	0.19	0.16	0.12
Relative Thickness	31%	27%	24%	22%	20%	19%	18%	17%	16%	15%	14%	13%	12%
Twist from Tip, °	8.8	8.1	7.3	6.6	5.9	5.1	4.4	3.7	2.9	2.2	1.5	0.7	0.0

Fig.3.3.4 Geometry of the AWEC-60 Rotor Blade

The rotor blade is a laminated fibre glass compound structure with polyester resin matrix. The main spar as load-carrying element along the full span of the blade is made of GFRP with approx. 80% of unidirectional glass-rovings. Being adapted to the shape of the leading edge of the profile, the section is of closed thin shell. The manufacturing process is a conformation of hand-lay-up laminate by means of an inner pressurised bag. The weight is about 3,500kg.

The trailing edge is built as a secondary structure (the so-called "afterbody panels"), made of a sandwich type of GFRP sheets and PVC foam. The supporting ribs are manufactured in GFRP. The total weight of the secondary structure is 600kg.

The forged flange (steel St-52) weighs 1,900kg. With additional 500kg of bolts, nuts and washers, the TOTAL weight of one rotor blade is about 6,500kg.

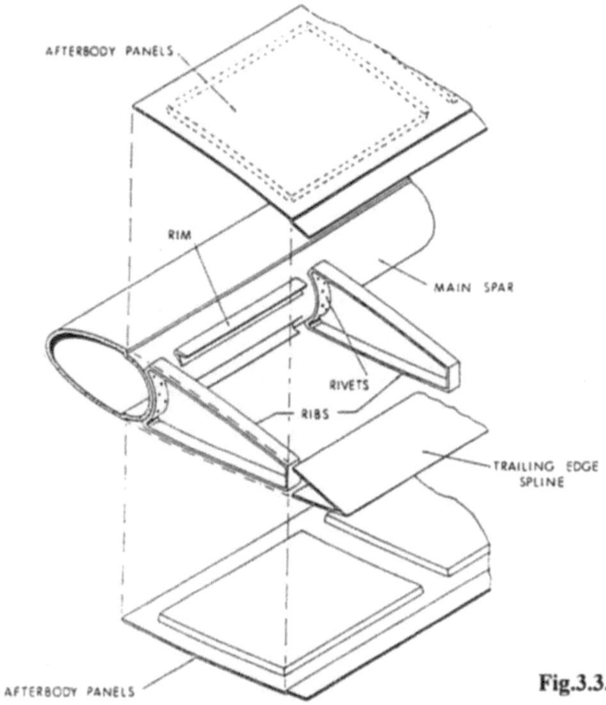

AFTERBODY PANELS

RIM

MAIN SPAR

RIVETS

RIBS

TRAILING EDGE
SPLINE

AFTERBODY PANELS

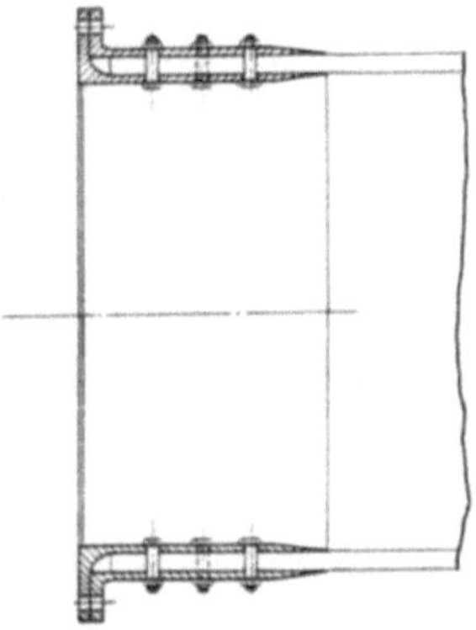

Fig.3.3.5 Design Concept of the AWEC-60
Rotor Blade and Steel Flange Attachement

Fig.3.3.6 Manufacturing of a Rotor Blade

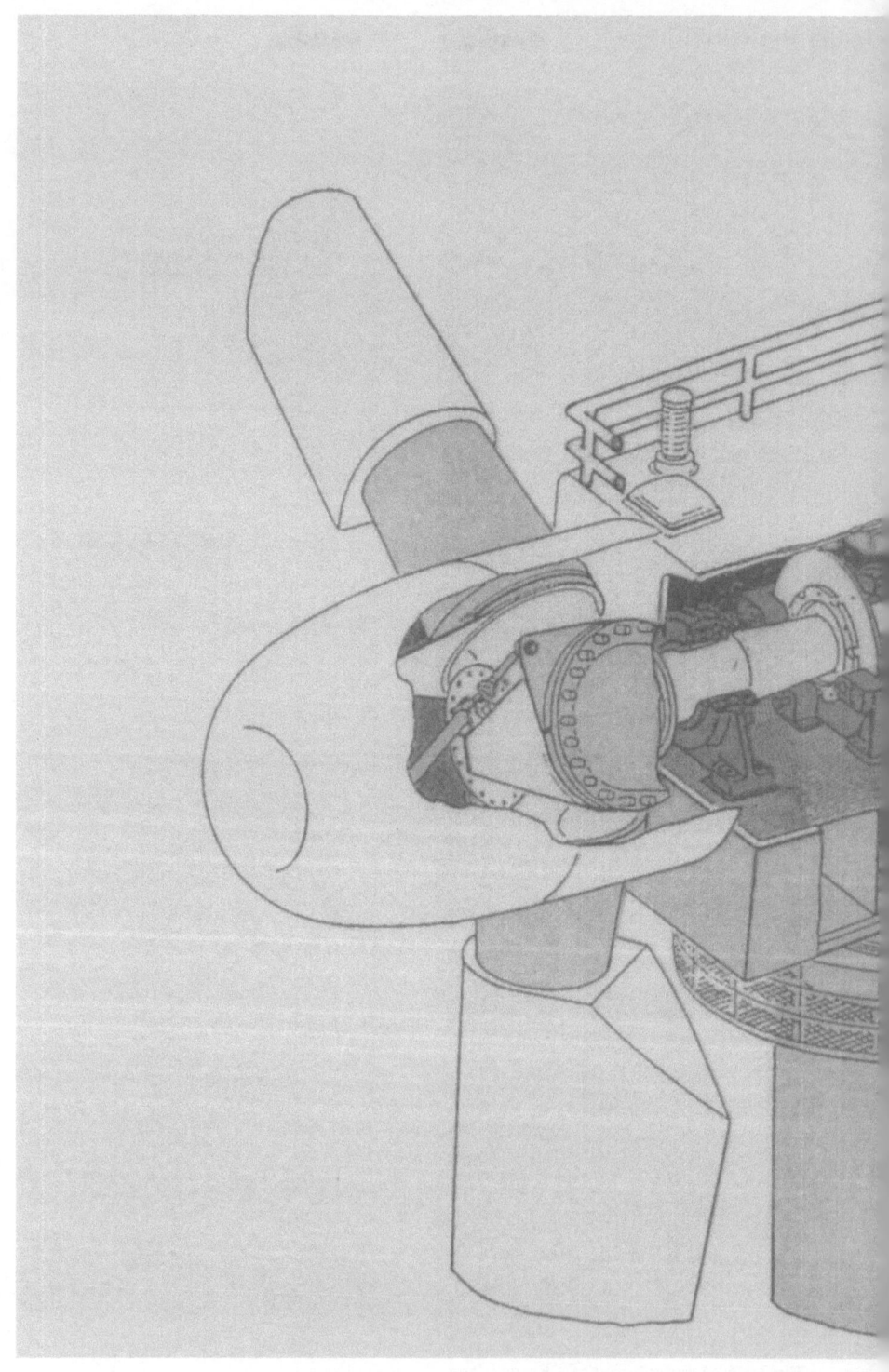

Fig.3.3.7 Artists View at the Nacelle of AWEC-60

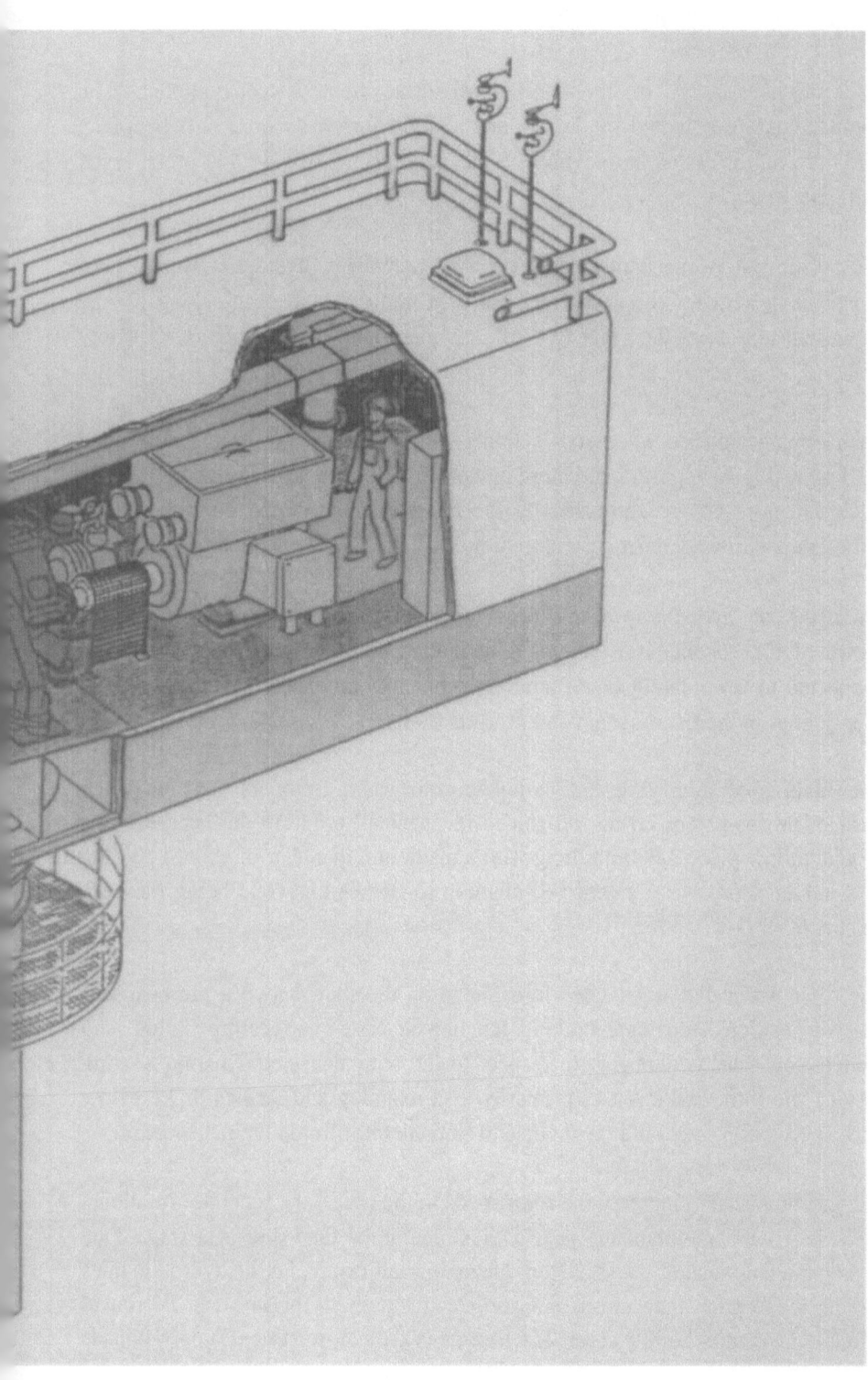

3.3.3 Mechanical Drive Train

The design concept of the mechanical drive train is conventional. The components are mounted modularly and straight forwardly on a stiff bedplate. The rotor hub is casted in one piece. Blade bearings are of a 4-point pre-stressed ball type.

The blade pitch mechanism relies on a redundant hydraulic system. An independent actuator for each rotor blade guarantees that the blocking of one blade will not affect the other two - this is important for the rotor brake safety system.

Each actuator consists of a servo cylinder with an emergency store. These units are fastened to a supporting structure mounted on the hub. The "individual" blade pitch actuators are synchronised by the microprocessor-based control system. Different positions will trigger a stop of the turbine. The blade pitch rate is 8°/s.

The necessary hydraulic power is provided by a stationary supply system in the centre of the nacelle. It is mounted in one rack for easy service works and connected to the rotating blade pitch adjustment components by two hydraulic supply lines in the hollow main and gearbox shafts.

The main shaft, carrying the complete rotor (hub, rotor blades and pitch mechanism) and transferring the gained mechanical torque to the gearbox, is a forged hollow piece 3,600mm long with a maximum diameter of about 850 mm. It is mounted on two standard self-aligning roller bearings, one being movable and the other one fixed.

The rotor brake disc is attached to a flange on the shaft between the bearings. This disposition keeps safe braking function in case of a gearbox failure and reduces the gear loadings. With respect to the redundant aerodynamic braking system, the rotor brake serves primarily as a retaining and service brake below rotor speeds of 3 rpm, but it is not able to stop the rotor from nominal speed.

The gearbox used to increase the rotor speed to the necessary generator rotational speed is a two-stage planetary gear with an additional final spur gear stage. Due to that, the output shaft is offset from the main shaft centre. So the hydraulic lines for the blade pitch mechanism and the electric lines to the hub can be routed through the central hollow shaft. The linkage to the non-rotational supply-units is made by a rotary duct for the electric lines and a slip ring arrangement for the hydraulic lines.

The gearbox is supported by two Teflon-coated bearing rings fixed to the base frame. The reactive torque is braced by two damped torque supports. This type of bearing mounting guarantees both a certain flexibility and an additional insulation against solid-borne noise.

A "rounded teeth" coupling is used on the main shaft side of the gearbox, and a double cardan coupling on the other side. At this end a hydraulic clutch is installed, serving as a safety measure in case of overloads.

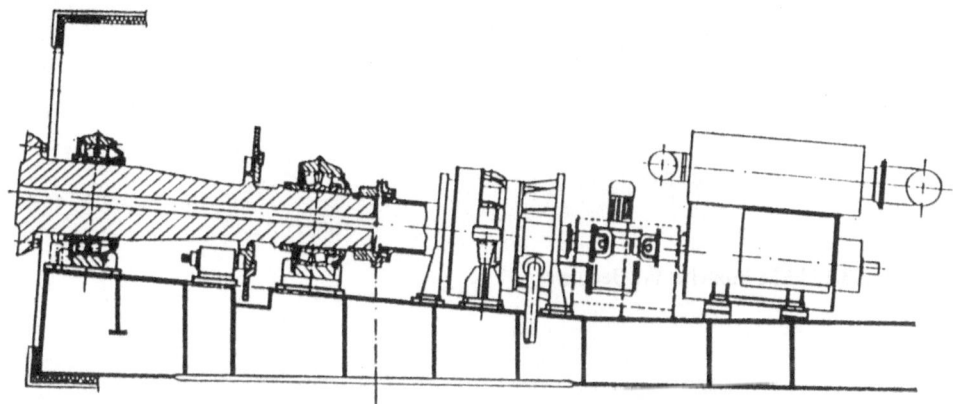

Fig.3.3.8 Mechanical Drive Train Arrangement

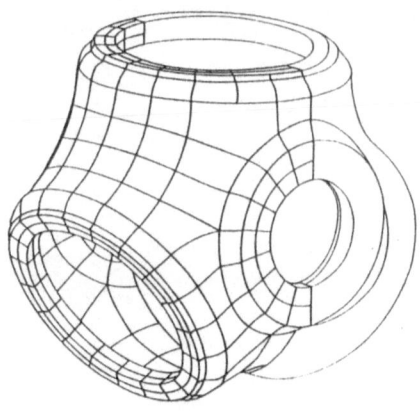

Fig.3.3.9 Rotor Hub: FEM Model

3.3.4 Nacelle Structure and Yaw System

The nacelle structure consists basically of two elements: bedplate and fairing. The bedplate is the structural base for the mechanical drive train on one side and for the yaw bearing on the other side. Due to this, a stiff geometry similar to a box girder was adopted; its size is 12.5 x 4.2 x 1.2m. The plates are in general 15mm thick and 20mm in the most loaded areas. At the yaw bearing attachment, a 65mm thick machined steel plate of very high characteristics is used. All welded areas are exposed for easy inspection during operative live.

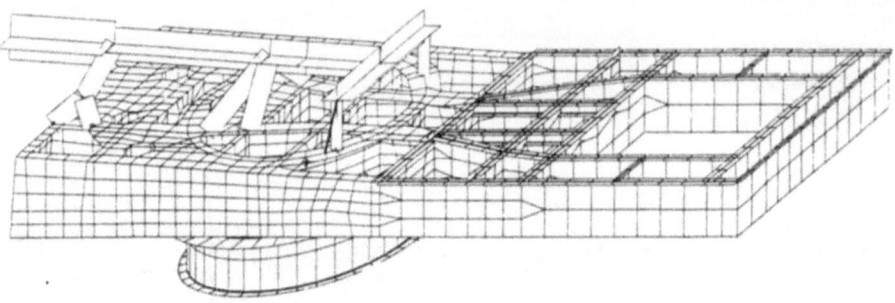

Fig.3.3.10 FEM-Model of the Bedplate

Fig.3.3.11 Welding of the Bedplate

The nacelle fairing located on top of the bedplate is designed with regards especially to good thermal insulation and effective reduction of noise emission. It is stiffed by a framework structure. The protection against corrosion of the nacelle was subject to intensive analysis, and a very severe specification was issued for the metal coating (aiming for 20 years of operational life with very minor maintenance even in high salinity ambience).

As yaw bearing, a standard bearing with automatic greasing is installed. A machined internal gear is integrated in the lower, tower-fixed ring of this bearing. The yaw drive is performed by the action of two travelling pinions using this gear. They are driven by an electric motor through gearbox and hydraulic coupling (all elements being fixed to the nacelle).

Yaw braking is provided by a set of 4 callipers also travelling with the nacelle and acting progressively on an annular track. A braking torque over 2,200kNm is obtained with this arrangement.

Fig.3.3.12 Nacelle during Assembly

3.3.5 Electrical System and Control

As the first of its kind on a large wind turbine, an innovative speed variable generator system is installed. The basic difference between the AWEC-60 system and a typical induction machinery used in other wind turbines is the introduction of a so-called "solid state oversynchronous cascade" linking the rotor rings to the grid.

This disposition allows controlling the rotor current and enlarges the usable slip range to +20%. The feed back of slip power leads to an increase in efficiency; moreover there is the possibility to "store" excess power in the rotating components through increasing the speed. This guarantees a smooth power delivery and contributes to alleviate loads and wear in the mechanical parts (like pitch mechanism and - specifically - the gearbox).

Additionally to the oversynchronous cascade there are three external electrical resistors which can be connected to the rotor of the induction generator instead of the cascade. Three different operating modes are possible:

- Oversynchronous cascade in operation: Speed variable mode
- External switchable resistor connected: "Flexible speed" mode with different generator slip.
- Generator short-circuited: Conventional fixed speed mode

The generator is an asynchronous machine with a rated power of 1,200kW and a corresponding nominal speed of 1,013rpm.

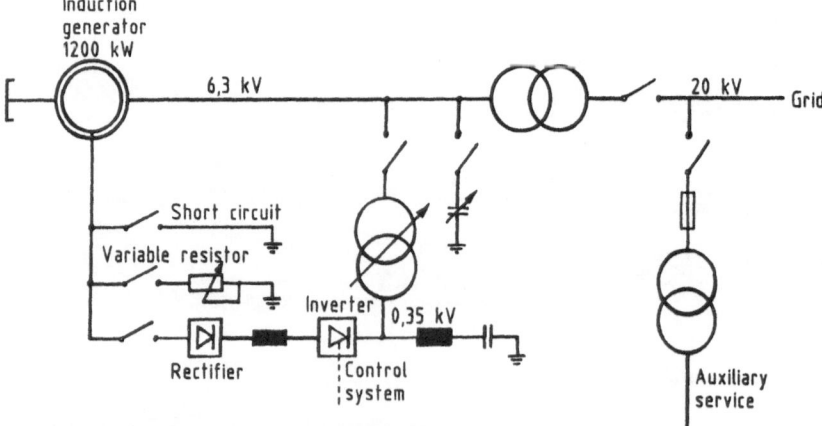

Fig.3.3.13 Electrical System of AWEC-60

The control system is based on a microprocessor in the nacelle and allows to optimise plant function and power output by modifying the software. It maintains the basic control - that are the steps of operational sequence - as well as the control loops for speed, power and yaw movement.

In general, the wind turbine operates automatically with the opportunity of an operator control from the operator's room in the tower base or from the supervising power plant. For this, a modem is installed to transfer commands and information via telecontrol by means of a phone line.

There are two different set points during normal operation - depending on the power output:

-   If the power output is below 1,200kW, the generator speed is stabilised at 1,100rpm through torque control; the power generation follows the action of wind speeds.

-   If the power tends to be higher than 1,200 kW, rotor slip and pitch angle are controlled to keep the power limited.

In case of a failure of the oversynchronous cascade, this would be disconnected, and the resistors instantaneously connected. A second failure in the resistors would lead to a short-circuiting of the rotor.

The equipment has a "self-protection" capacity which takes the decision of connection or disconnection in case of detection of failures or after checking that working parameters are out of the operative range (e.g. different blade pitch angles, overspeed, vibrations, pressure loss etc.).

3.3.6 Tower and Foundation

The free-standing tower of AWEC-60 is made of steel ST-52, 20mm thick, except reinforced areas at yaw drive and the cone-cylinder transition. The cylindrical part has a diameter of 3.5m and connects to a truncated cone 12m high with a 8m diameter at its lower base. The conical part improves the stiffness of the tower at the most loaded sections, and gives space for the installation of conventional electrical equipment.

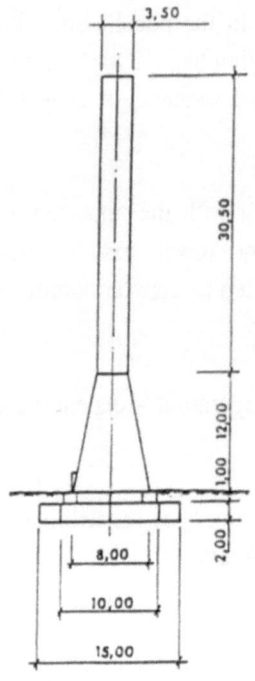

The foundation basically consists of a reinforced concrete slab of octagonal shape (characteristic diameter 15m) with a total weight of 1,100tons. The height of the slab differs between 2 and 3m. The conservative design adopted leads to small loads on the ground and high safety factors both for overturning moment and stress appearing in the concrete.

Fig.3.3.14 Tower and Foundation of AWEC-60 at Cabo Villano

Fig.3.3.15 Construction of the Foundation at Cabo Villano

3.3.7 Assembly, Transportation and Erection

Due to the remote site at Cabo Villano the transportation and erection of the large wind turbine have been major items. Three large main components of the turbine were manufactured and transported to Cabo Villano independent from each other: tower, nacelle and rotor blades.

The tower was land-transported: the whole cylindrical part in one piece 31m long and with a weight of 54tons, and several sectors of the base cone. They were welded together on the ground and then erected by a lift with a huge portic synchronised with a horizontal sliding on Teflon pads. This portic was a welded structure with a 60m high column and a 14m long girder beam.

The nacelle was assembled in two stages: First in Madrid at BOETICHER Y NAVARRO S. A. Workshop, the main components of the drive train such as shaft, bearings and gearbox together with the central hydraulic oil supply system were mounted on the bedplate of the nacelle. Then, this unit was road transported to LA CORUNA.

Secondly the missing elements of the drive train and the pitch mechanism, the control and data acquisition system, the nacelle fairing and the yaw system have been assembled in another workshop at LA CORUNA.

Extensive tests of all functions of the completed nacelle including the yaw system finished the workshop activities. The simulation of the actual working conditions of the plant allowed the amending of minor faults. After that, this unit with a weight of about 180tons and with outer dimensions of 16.2 x 4.5 x 5.0m was shipped (due to the road conditions) to the port of CABO VILLANO. At last it needed 5km of road-transportation to the wind turbines site.

The three rotor blades were brought to the site on a tractor with a 3-axle platform. This was necessary to provide some bracket points to avoid buckling problems.

The nacelle was lifted by a slightly tilted portic: A rotation of it effected the horizontal moving over the top of the tower. The rotor blades were mounted separately in horizontal position by two cranes.

Fig.3.3.16 Transport of the Nacelle

Fig.3.3.17 Transport of a Rotor Blade

Fig.3.3.18 Lifting of the Nacelle

Fig.3.3.19 Mounting of a Rotor Blade

4

Technical and Cost Comparisons

A comprehensive technical comparison of the WEGA wind turbines is the objective of the Design Review and Evaluation (see chapter 2). The results of this programme will not be available before 1994. However, it is possible to discuss and compare essential technical criteria from the conceptual point of view. This is done here because they are important to the understanding of the technology.

4.1   Innovative Design Elements

The overall technical concept of the three WEGA machines conforms to that of most of the today's commercially available wind turbines in the power-range up to 500kW. The main features of this conventional concept are: three blade upwind rotor, modular drivetrain mounted on a bedplate and an electrical system based on conventional electric generators.

On the subsystem and component level some innovative features have been applied to the WEGA projects. The innovative design elements are highlighted in figure 4.1 and the technical configurations of the all three turbines are compared. Innovative effort is concentrated upon the rotor and upon the power control and electrical system. The drive train concepts are all similar as are the load carrying structures in the nacelle; all use a conventional bedplate. This latter observation is interesting because drivetrain layout and bedplate design is an area of rapid innovation in the smaller commercial machines today.

	TJAEREBORG 2MW	*RICHBOROUGH 1MW*	*AWEC-60 1.2MW*
Rotor Blades:			
Design/Material	glass fibre/polyester steel flange	wood/epoxy, no steel flange, thick airfoils	glass fibre/polyester steel flange
Manufacturing	mechanical winding of loadcarrying spar	hand laminated	hand-lay-up formed by inner-pressurised bag
Mechanical Drivetrain:	modular, beplate mounted brake on high speed shaft	modular, beplate mounted brake on low speed shaft	modular, beplate mounted brake on low speed shaft
Power Control: and	full span pitch	partial span pitch	full span pitch
Electrical System:	induction generator with enlarged slip (2%)	induction generator	induction generator & oversynchr.cascade for variable rotor speed, 3 operational modes
	fixed rotor speed	fixed rotor speed	
Safety System:			
Aerodynamical	each blade redundant hydraulic pitch system	each blade redundant hydraulic pitch system	each blade redundant hydraulic pitch system
Mechanical	hydraulic disc brake	hydraulic disc brake	hydraulic disc brake
Nacelle:	very large in respect to the R&D-character	very compact "value engineered"	large in respect to the R&D-character
Tower:	tapered concrete tower	tubular steel tower conical base	tubular steel tower conical base

Fig.4.1 Comparison of the Main Subsystems and Components of the WEGA Wind Turbines with the innovative features as indicated

4.2 Structural Stiffness

The design of some essential components of a wind turbine, i.e. the tower, is dominated by the stiffness requirements to avoid dangerous resonances with the rotating blades. Therefore the approach to structural stiffness in the design of a wind turbine can have a very important bearing on the component weights and as a result of this the production costs. Furthermore the flexibility of the structure affects the response to the dynamic loads caused by the air turbulence and this can also be important.

The main technical criteria which govern the structural stiffness concept are the bending eigenfrequencies of those components which can be exited by the rotational frequency of the rotor and its higher harmonics. In the case of the three bladed WEGA turbines the 'three per revolution' (3P) excitation of the rotor is the main source of a possible resonance oscillation. Other harmonics of the blade rotation frequency can also have an influence but they are minor in comparison with the blade passing frequency.

The most important eigenfrequency values of the WEGA wind turbines are summarised in figure 4.2. The rotor blade eigenfrequencies are as follows:

-        1. flap: 1st bending eigenfrequency out of rotor plan
-        1. lag: 1st bending eigenfrequency in rotor plan
-        2. flap: 2nd bending eigenfrequency out of rotor plan

In each case the first bending eigenfrequencies of the towers are the lowest resonances in the system in both the axial and the transverse direction. Due to the dynamic behaviour of the rotating masses the eigenfrequencies are somewhat higher in the rotating system than in the static system.

[Hz]	TJAEREBORG 2MW		RICHBOROUGH 1MW		AWEC-60 1.2MW	
	System static	System rotating	System static	System rotating	System static	System rotating
Blade Passing Frequency (3P)	-	1.12	-	1.23	-	1.03...1.27
Rotor Blades:						
1. flap	1.10	1.24	1.34	1.42	1.31	1.39
1. lag	2.30	2.34	2.44	2.57	2.58	2.76
2. flap	3.22	3.34			3.87	4.28
Tower Bending:	0.78	0.81	1.07	1.08	0.75	0.77

Fig.4.2 Main Bending Eigenfrequencies of Rotor Blade and Tower

Figure 4.3 shows the resonance diagrams for the three WEGA turbines. It is obvious that the stiffness concepts are very similar. The 1st rotor blade eigenfrequency is above the 3P excitation, whereas the 1st bending eigenfrequency of the tower is between the 2P and 3P lines.

The value of the 1st bending eigenfrequency dominates the tower weight; this is particularly so for tubular steel towers. In very recent developments the tendency has been towards 'super soft' towers, with the 1st bending eigenfrequency below the 1P excitation.

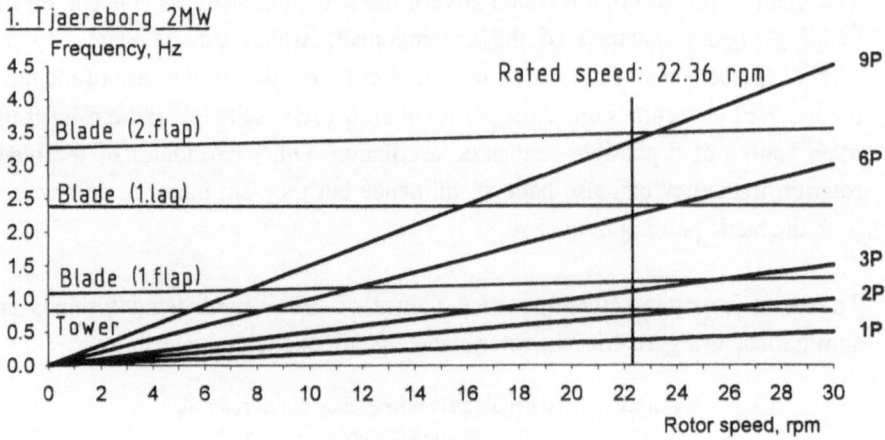

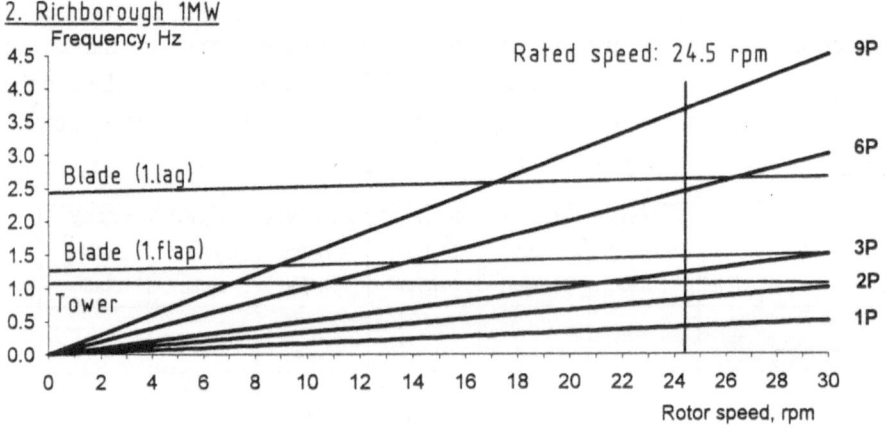

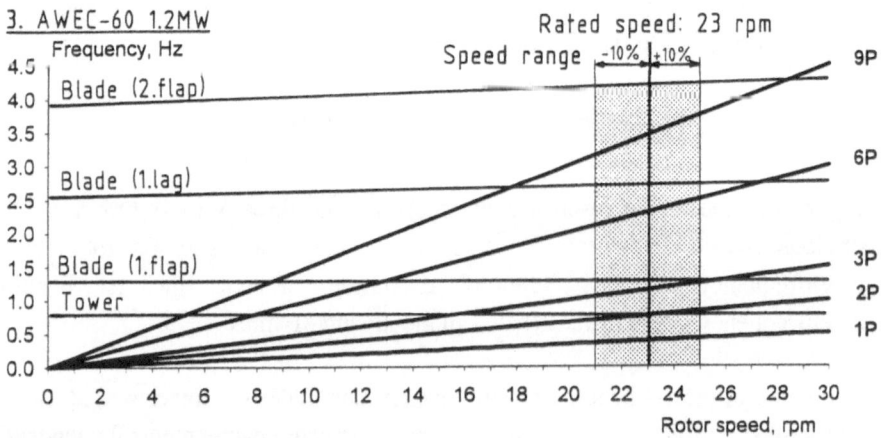

Fig.4.3 Resonance Diagrams of the WEGA Wind Turbines

4.3   Component Weight Breakdown

The detailed component weight breakdowns of the WEGA turbines are presented in figure 4.4. While the Tjaereborg wind turbine and the AWEC-60 only show interesting differences in detail, the Richborough wind turbine design is obviously considerably more lightweight.

The specific towerhead weight of the Richborough turbine is only 35kg/m² (weight per rotor swept area), compared to 65...75kg/m² for the Tjaereborg and AWEC-60 turbines (Fig.4.5 and Fig.4.6). This reflects the fact that the Richborough turbine has been developed as part of a commercial production series, stepwise from 250kW to 750kW and finally to 1,000kW, with the result that some degree of 'value engineering' has been included in the design process. Thus the Richborough turbine uses the same technical concept as the 750kW unit and the design of some of the components are identical.

The very high specific weight figures of the Tjaereborg wind turbine and the AWEC-60 are a consequence of the R&D character of these projects. The main centres of the high weights are (compare Fig.4.7):

-    heavy rotor blades

-    extremely long mechanical drive train arrangements particularly the rotor shaft and bearings

-    very high volume nacelles, i.e. at the Tjaereborg machine

The specific towerhead weights of the WEGA turbines are compared against the background of the today's commercial wind turbines and some other experimental, large wind turbines in figure 4.8. The AWEC-60 and especially the Tjaereborg turbine are in the upper range of the weights which are given by an extrapolation from 'classical' Danish concepts of the early eighties. The Richborough machine fits into the range of the today's state of the art commercial three bladed turbines.

Description:		TJAEREBORG 2MW	RICHBOROUGH 1MW	AWEC-60 1.2MW
Rotor Diameter	[m]	61	55	60
Swept Area	[m²]	2,932	2,376	2,827
Number of Blades	[-]	3	3	3
Rated Power	[kW]	2,000	1,000	1,200

Weights:	[kg]			
Rotorblade (each)		9,033	3,600	6,633
Σ:	**Rotorblades**	**27,099**	**10,800**	**19,899**
Hub Structure		22,100	10,500	25,600
Blade Pitch Mechanism		in "Auxiliary Systems"	in "Auxiliary Systems"	in "Aux. Systems" (9.000)
Low Speed Shaft		21,000	3,750	25,400
Low Speed Shaft Bearings		9,400	2,270	in "... Shaft"
Gearbox		16,200	9,500	11,700
Brake		in "Auxiliary Systems"	in "Auxiliary Systems"	in "Aux. Systems" (4.500)
High Speed Shaft		in "Auxiliary Systems"	in "Auxiliary Systems"	in "Aux. Systems" (300)
Generator		7,800	4,500	7,700
Σ:	**Mechanical Drivetrain**	**76,500**	**30,520**	**70,400**
Bedplate		21,600	16,500	49,000
Nacelle Fairing		34,000	4,000	14,000
Spinner		in "Auxiliary Systems"	none	in "Hub Structure"
Σ:	Nacelle Structure	55,600	20,500	63,000
Yaw System		in "Auxiliary Systems"	in "Auxiliary Systems"	in "Aux. Systems" (11.000)
Hydraulic Supply System		in "Auxiliary Systems"	in "Auxiliary Systems"	in "Aux. Systems" (3.000)
Electrical Equipment		in "Auxiliary Systems"	in "Auxiliary Systems"	in "Aux. Systems" (2.000)
Auxiliary Systems in Nacelle		65,400	21,580	36,300
Σ:	**Nacelle**	**121,000**	**42,080**	**99,300**
ΣΣ:	**Towerhead**	**224,599**	**83,400**	**189,599**
	Tower	645,000	75,000	92,000
	Equipment placed in Tower	20,000	1,500	in "Tower"
ΣΣΣ:	**Over All**	**889,599**	**159,900**	**281,599**

Specific Weights:				
Rotorblades/Swept Area	[kg/m²]	9.2	4.5	7.0
Mech.Drivetrain/Swept Area	[kg/m²]	26.1	12.8	24.9
Nacelle/Swept Area	[kg/m²]	41.3	17.7	35.1
Σ: Towerhead/Swept Area	**[kg/m²]**	**76.6**	**35.1**	**67.1**
Towerhead/Rated Power	[kg/kW]	112.3	83.4	158.0

Fig.4.4 Weight Breakdown of the WEGA Wind Turbines

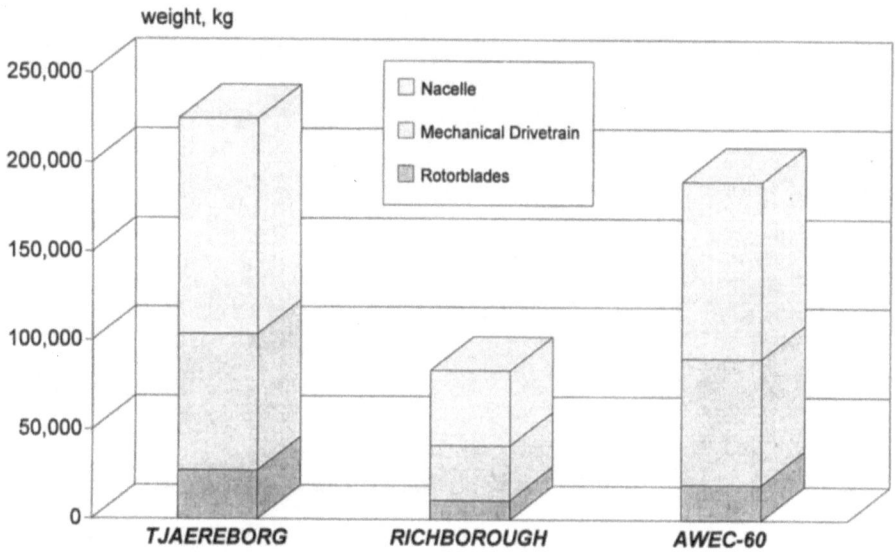

Fig.4.5 Weights of the Main Towerhead Subsystems

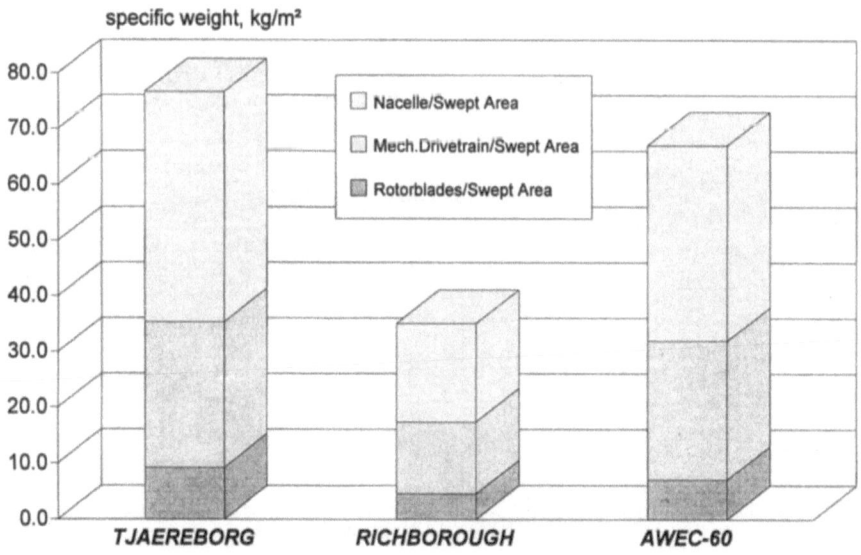

Fig.4.6 Specific Towerhead Weights

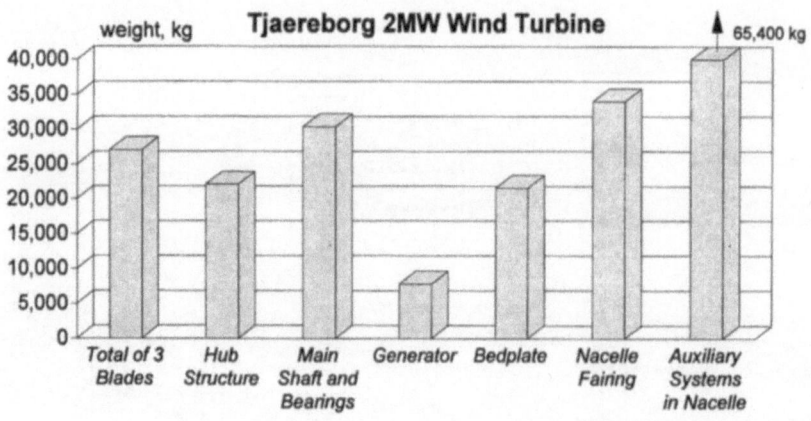

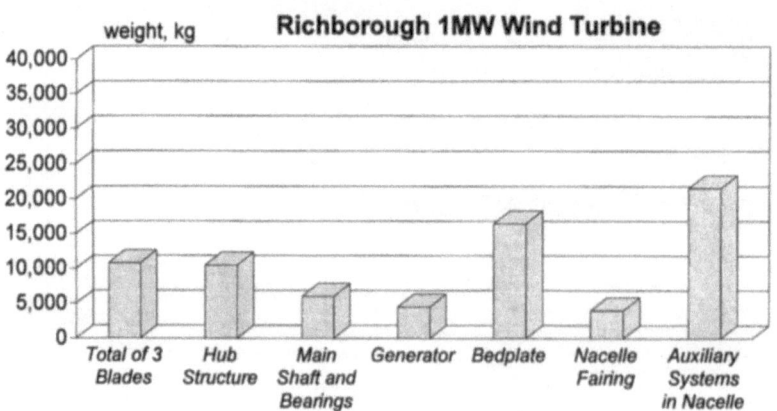

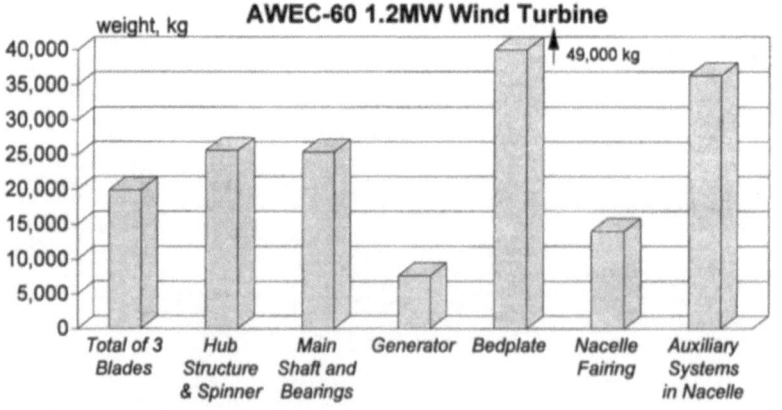

Fig.4.7 Comparison of Component Weights for the three WEGA Turbines

Specific Towerhead Mass kg/m²

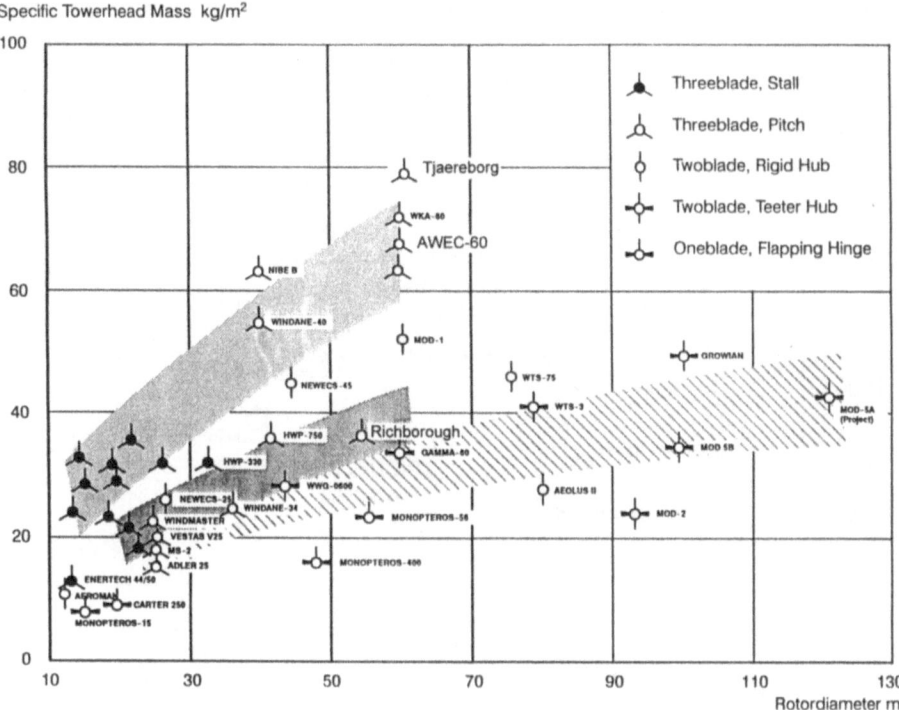

Fig.4.8 Specific Towerhead Weights of Existing Wind Turbines /9/

4.4   Manufacturing Cost

Cost comparisons in the framework of R&D projects are very difficult. The development costs cannot be compared in a consistent way and depend on the developer's preconditions and, also, on his particular cost structure to a large extent. Furthermore overruns of the official budgets often are not published.

The manufacturing costs of the prototypes are a more reliable indicator of the economic potential of the concept, but they cannot be separated very clearly from the R&D expenses.

However, the expected ranges of the manufacturing costs of the WEGA turbines are indicated in figure 4.9 in terms of specific values ECU/m². Because of its lightweight design the Richborough wind turbine is in the lower range of the experimental wind turbines' manufacturing costs, whereas the Tjaereborg turbine and the AWEC-60 both lie in the upper range.

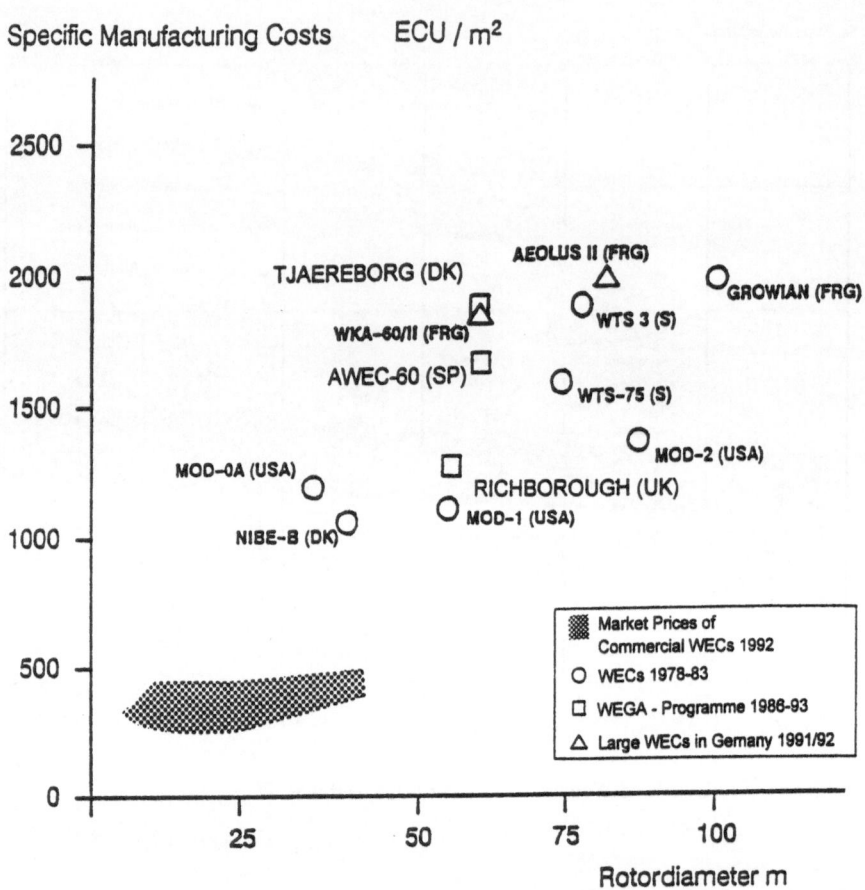

Fig.4.9 Specific Manufacturing Costs per Rotor Area for Commercial and Experimental Large Wind Turbines (Costs referring to the year of construction)/   /

It is obvious that all of the large wind turbine which have been built up to now have specific costs which are far above the market prices of commercial wind energy converters. Even a series production process giving a reduction in manufacturing costs of 30% up to 40% would not meet the commercial goal. To achieve this, substantial improvements in the design of the main structural components are necessary to reduce towerhead weights.

The proportions of the cost breakdown attributed to the main components do not differ significantly from those of today's commercial machines (Fig.4.10). This indicates that it is not sufficient to concentrate the cost reductions on a single component. The whole system has to be optimised in this respect.

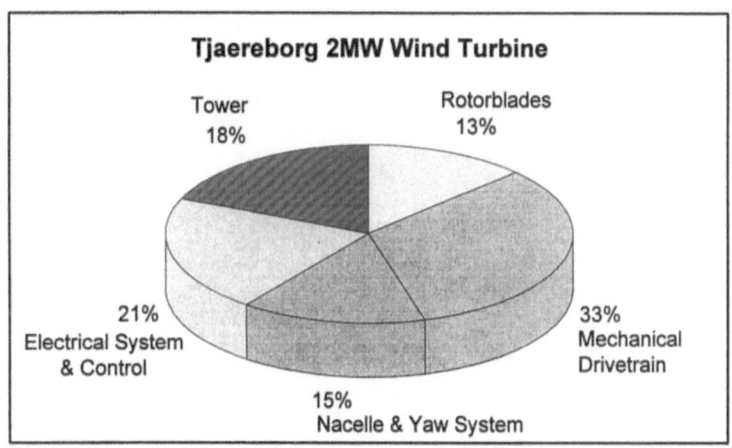

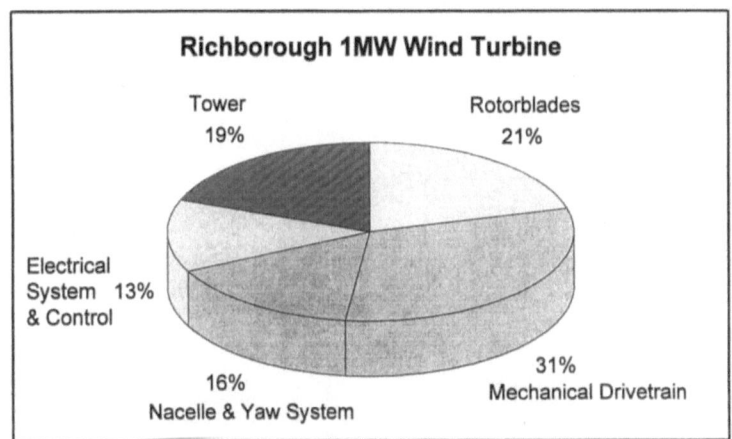

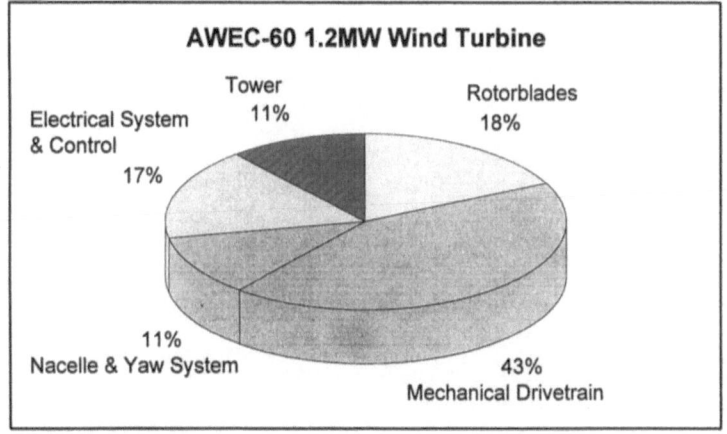

Fig.4.10 Manufacturing Cost Breakdown to the Main Subsystems

5
Commissioning and Early Operational Experience

The evaluation of the WEGA wind turbines is not expected to show commercial standards of operation. The main emphasis is on the R+D work. Nevertheless the operational records of so far suffered from the fact that each turbine has had a long outage period due to a major component failure. In the case of the Richborough turbine and the AWEC-60 changes in the organisations responsible for the projects and technical problems with the data acquisition systems have caused further delays. This background has to be considered when judging the early operational experiences.

5.1 Commissioning and Acceptance Tests

The essential pre-condition for the commissioning of a wind turbine is the first successful connection of the generator to the grid. Furthermore all operational sequences, i.e. safety test have to be performed, before the wind turbine is ready for automatic operation.

To ensure that the manufacturing and the commissioning phase of the project have been duly carried out, an acceptance test scrutinised by representatives of the Commission was required for the WEGA wind turbines. A detailed programme for the acceptance test was worked out by the contractor and approved by the Commission. This programme had to demonstrate that, upon completion of the acceptance test, the wind turbine is ready and available for unmanned operation according to the final design requirements.

As a minimum the acceptance tests had to comprise the following elements:

a) Without grid connection.
    - overrun with check of overrun protection and stop procedure.
    - start up with check of start-up procedures.
    - automatic stop with check of stop sequences.

b) Grid connection.

c) 36 hours of unmanned operation at wind speeds higher than cut-in wind
    speed.

d) Transient testing.
    - Loss of grid.
    - Generator trip.
    - Cut out for high wind, possibly "artificial high wind".
    - Emergency stop from low and high wind speed.

In the time table below the dates of commissioning, i.e. first connection of
generator to grid, and acceptance tests for the three WEGA wind turbines are
indicated.

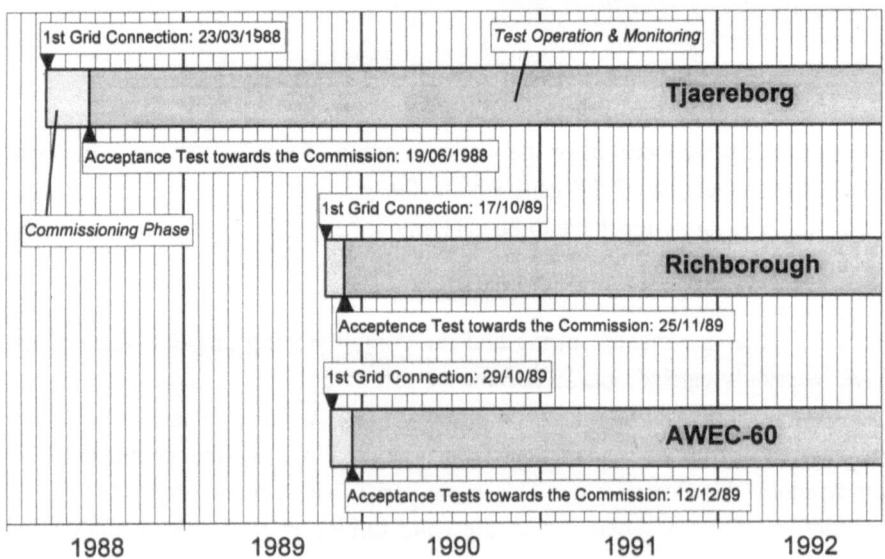

Fig.5.1 Dates of Commissioning Phase and Acceptance Tests of the WEGA Wind Turbines

Tjaereborg Wind Turbine

On March 23, 1988, the generator was connected to the grid for the first time. The acceptance test to meet the requirements of the CEC, DG XII, commenced on Wednesday June 15, 1988 and was completed on Sunday June 19, 1988 with the exception of a single high wind cut out test which could not be performed due to low wind speeds. The remaining test was performed on September 2, 1988 and the required test report for the Commission was approved.

Typical time histories showing power, pitch angle and wind speed at hub height at a power level of approx. 2000kW are shown in figure 5.2. The change between fixed pitch operation and pitch regulation for power control can clearly be seen as well as the 3P harmonic in the power signal caused by the blades passing the tower. The apparent lack of correlation between wind speed, pitch and power is due to the wind blowing abeam of the meteorological tower and the wind turbine.

Richborough Wind Turbine

The generator was first synchronised on 17th October 1989 and power output reached 500kW the following day. The series acceptance tests were defined and the majority were performed in late November 1989. All of those undertaken were successfully completed.

The final test - a 36 hour unmanned run - was postponed until one or two teething problems had been overcome and wind conditions allowed running at high output power levels. This final test was satisfactorily completed on 2 May 1990.

Since May 1990 the machine has been running in automatic mode with only occasional stops for instrument calibration. The running hours by the end of August 1990 were 949 and the recorded output was 329.31MWh. On July 10th the turbine was officially opened by the Under Secretary of State for Energy, Mr. P. Morrison at a ceremony attended by officials of the CEC.

Figure 5.3 shows the variation of low speed shaft speed and torque during start up. It can be seen that the shaft speed increases rapidly until the control system starts to limit rotor speed. Final synchronisation occurs about 90 seconds after the rotor is released. The torque transients on synchronisation are well within the design limits for the shaft.

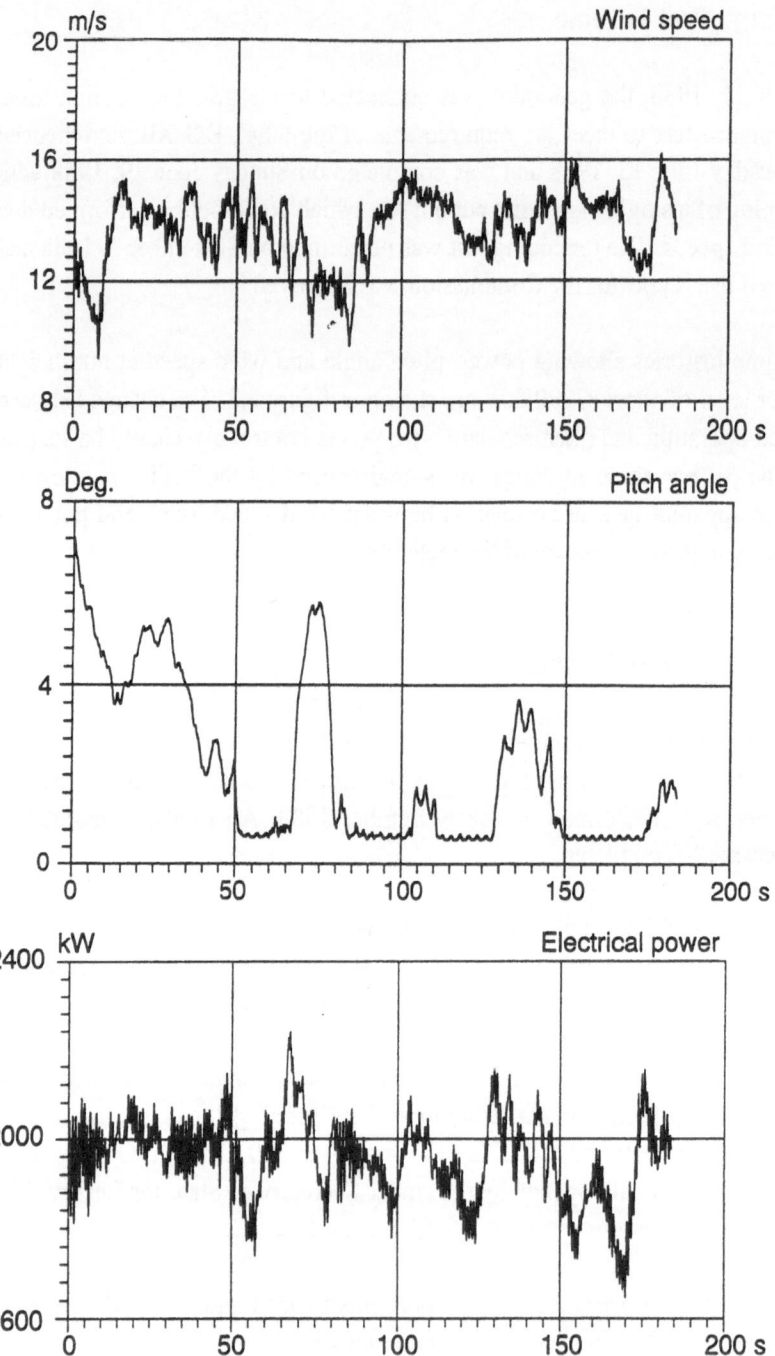

Fig.5.2 Typical Time Histories of Wind Speed, Pitch Angle and Power Output at the Tjaereborg Wind Turbine, Measured During the Commissioning Phase

Shown in figure 5.4 are just 10 seconds of wind speed and power generation data recorded at 40Hz. As might be expected there is little correlation between the two on this time scale (the wind speed measurements and the wind turbine are separated by about 200m). The power output is dominated by a regular oscillation of ±7% at the blade passing frequency which naturally results from operation in a sheared wind profile.

AWEC-60 Wind Turbine

On the dates 11 and 12 December 1989, representatives of the Commission, Asinel, Union Fenosa, IER-CIEMAT and MAN met at Cabo Villano to perform the contractual test for the acceptance of the AWEC-60 wind turbine.

This test was, in fact, a repetition of the commissioning tests which had been performed by Asinel in the presence of Union Fenosa on the dates 29 and 30 October 1989.

The test, which had been carried out in full accordance to the procedures established in the Contract, proved that the machine is capable of performing all the required operations.

As a typical example of the commissioning tests, figure 5.5, shows a start-up sequence of the AWEC-60 machine. At a wind speed of 10-15m/s the sequence starts with the pitch angle being 65° and the rotor idling at approximately 4.2rpm, as indicated by the 200rpm recorded at the high speed shaft. Then the pitch angle is diminished, leading to an acceleration of the rotor speed. When the high speed shaft has been running at a velocity of more than 1000rpm for a significant time, the generator is connected to the grid. With the pitch angle controlled down to 5-10° the rated rotor speed is reached and the machine generates a mean power of approximately 1000kW.

In figure 5.6 the same parameters are indicated for normal operation at a mean wind speed of approximately 9m/s. The variable speed operation of AWEC-60 can be seen in the velocity measured on the high speed shaft.

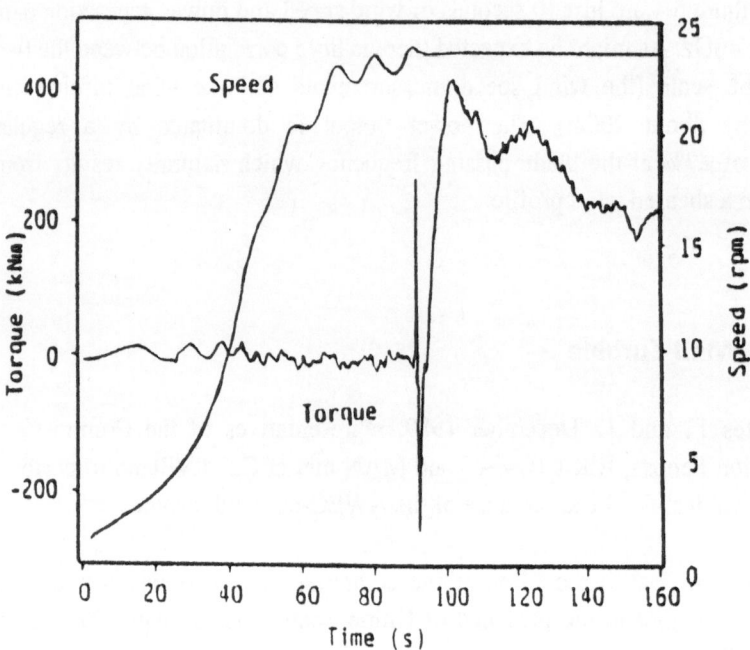

Fig.5.3 Low Speed Shaft Speed and Torque during Start-Up of the Richborough Wind Turbine

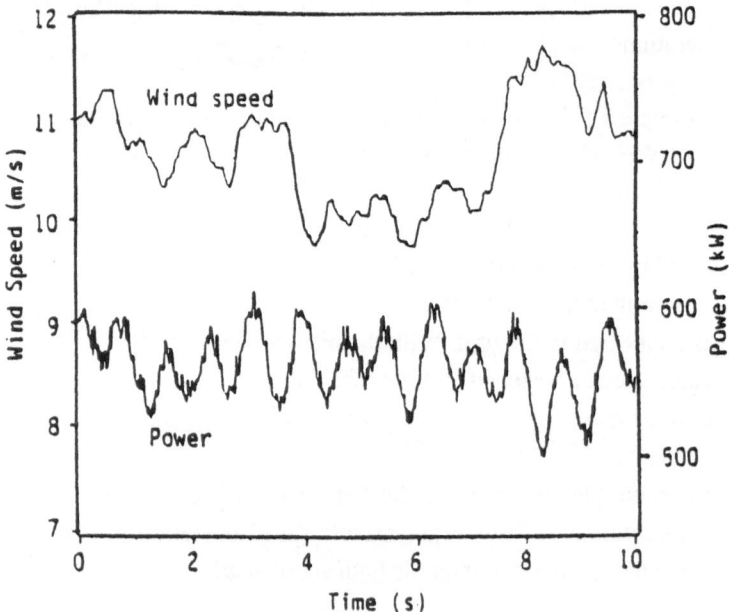

Fig.5.4 Time History of Wind Speed and Generated Power at the Richborough Wind Turbine, Measured During the Commissioning Phase

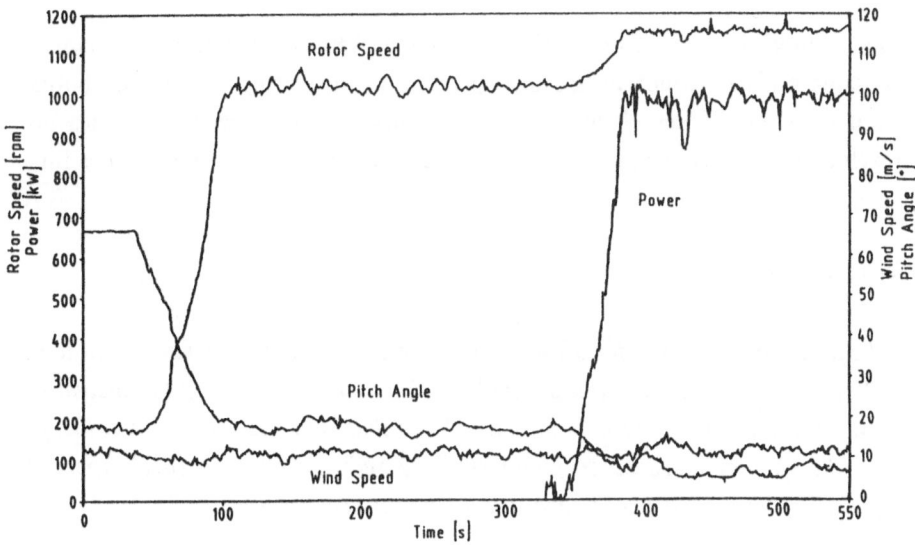

Fig.5.5 Time Histories of Rotor Speed, Wind Speed, Pitch Angle During a Start-Up Sequence up to Nearly Rated Power of the AWEC-60 Wind Turbine

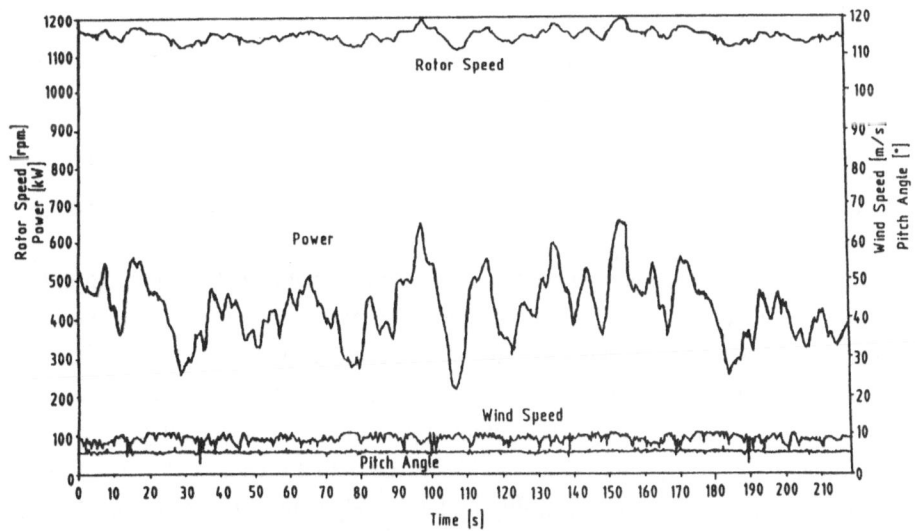

Fig.5.6 Time History of Wind Speed, Rotor Speed and Power of the AWEC-60 Wind Turbine at Partial Load Conditions (fixed blade pitch angle)

5.2   Measured Power Curves

The contract between CEC, DG XII and the WEGA wind turbine operators contains a clause which deems the operators to repay part of the contribution from the Commission, if the gross power curve of the wind turbine does not reach a certain level compared to the guaranteed power curve given in the final design report.

The contract says:

"The power curve values for all wind speeds greater than 6.5m/s must be at least 90% of the power curve values stipulated in the final design. If the measured power curve at any point is less than 90% of that stipulated in the final design then the Commission's contribution shall be reduced by (X - 10%) where X represents the largest percentage deviation between the measured power curve value and the power curve stipulated in the final design."

Tjaereborg Wind Turbine

The power curve was measured in the period from November 13, 1990, when the turbine was put in regular operation after a gearbox failure, until August 31, 1991.

The power curve is corrected to standard conditions in accordance with the IEA recommendations No.1 "Power Performance Testing", 2nd edition 1990 /3/.

Figure 5.7 shows the power curve together with the guaranteed curve given in the final design report. The measured power curve is considerably above the guaranteed curve for all wind speeds. The reason for this deviation is that the guaranteed curve is calculated for NACA standard rough profile data in order to take possible difficulties with obtaining smooth blade surfaces in the manufacturing process into account. The blade actually came out with smooth surfaces and hence with considerably higher aerodynamical efficiencies, and this explains the deviation.

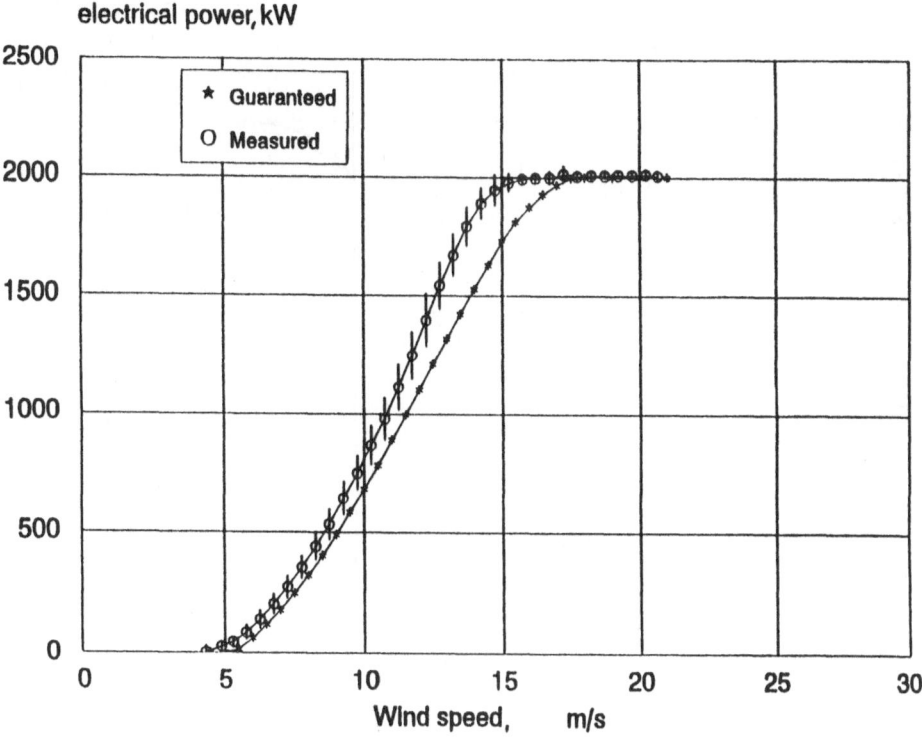

Fig. 5.7 Measured and Guaranteed Power Curve of the Tjaereborg Wind Turbine

Richborough Wind Turbine

The measured, density corrected power curve of the Richborough 1MW wind turbine is compared to the predicted curve given in the final design in figure 5.8. The measured power curve is below the guaranteed curve at lower wind speeds up to 11.5m/s whereas at higher wind speeds the generated power is higher than predicted.

AWEC-60

The power curve measurements and calculations for the AWEC-60 wind turbine were been carried out in the first quarter of the year 1991 in accordance with the IEA recommendations /3/. Figure 5.9 shows that the measured power curve lies above the predicted curve. The theoretical calculations of the power curve were proven to be very conservative.

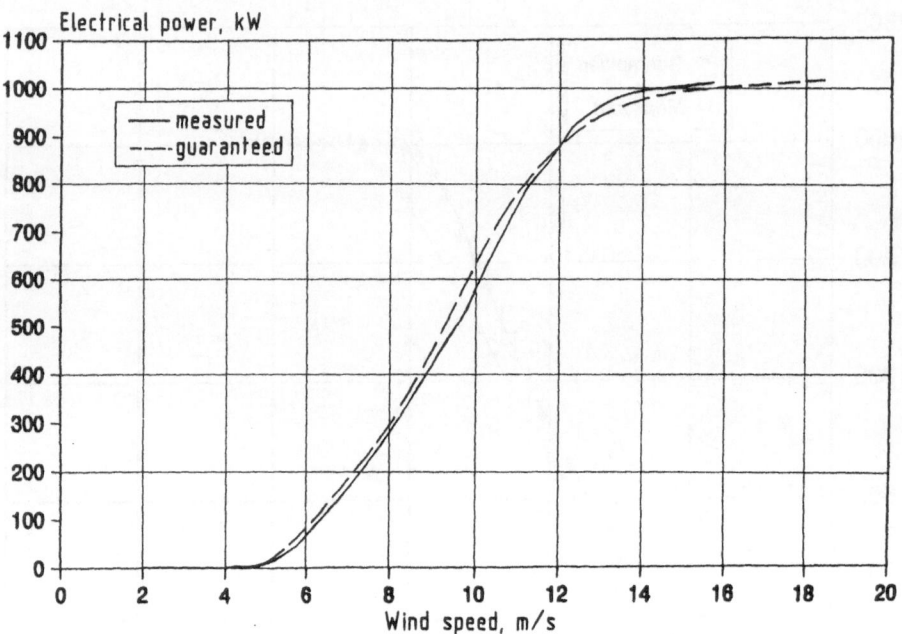

Fig.5.8 Measured and Guaranteed Power Curve of the Richborough Wind Turbine

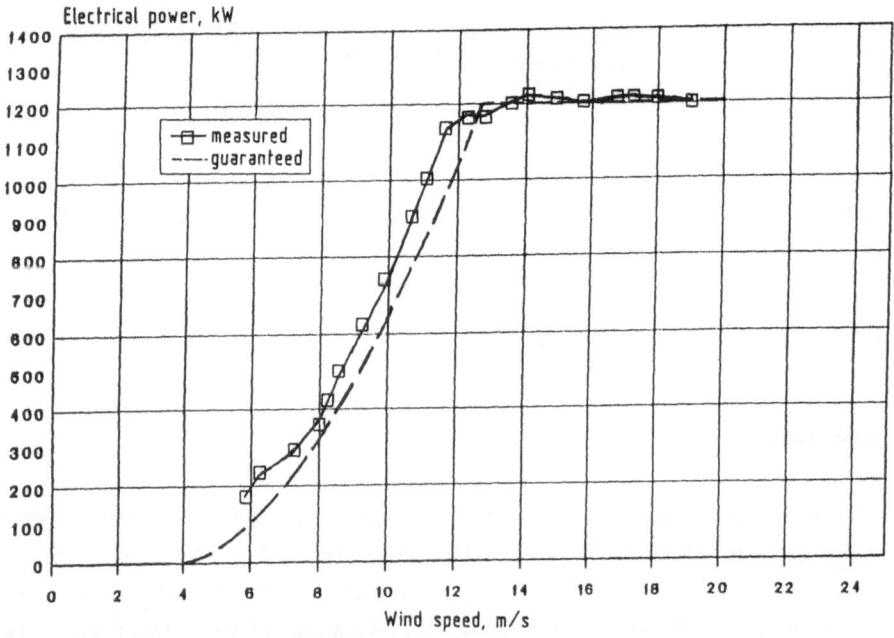

Fig. 5.9 Measured and Guaranteed Power Curve of the AWEC-60 1.2MW Wind Turbine

Comparison of the Measured Power Curves

The measured power curves of the three WEGA wind turbines are compared to each other in figure 5.10. The maximum power is determined by the rated power of the installed generator. In the partial load area differences caused by the rotor size, aerodynamic efficiency of the rotor blades and the influence of the rotational speed control are obvious.

The curves of the overall power coefficient c_p versus wind speed, given in figure 5.11, are calculated from the above mentioned measured power curves.

The Tjaereborg wind turbine shows the best power performance in the wind speed range above 9m/s. The measured overall power coefficient achieves its maximum of c_p=0.44 at a wind speed of 9.8m/s, which is the design wind speed of the turbine. Referring to the predicted maximal rotor power coefficient of $c_{p,R}$=0.485 the resulting combined mechanical and electrical losses are found to be approximately 9 to 10%.

Referring to the given c_p-values the aerodynamical efficiency of the Richborough wind turbine's rotor is significantly below those of the other two WEGA machines. Probable reasons for this may be the thicker airfoil section and the partial span control system, i.e. the slot between the inner blade area and the pitchable blade tips.

Because the variable speed control system, the power coefficient of the AWEC-60 wind turbine achieves the best values at wind speeds below 8m/s. The matching of the rotor speed to the wind speed keeps the tip speeds ratio in the range of the design tip speed ratio even at a wind speed of 6m/s. Hence the rotor performance of the AWEC-60 does not decrease as much as the fixed speed rotors of the Tjaereborg and Richborough turbines do at wind speeds below design wind speed.

As expected, the main advantage of the variable speed concept is proven. It gives higher energy production in the lower partial load area, leading to an increase in electricity production of 5 to 8% when compared to a fixed speed machine with the same rotor size.

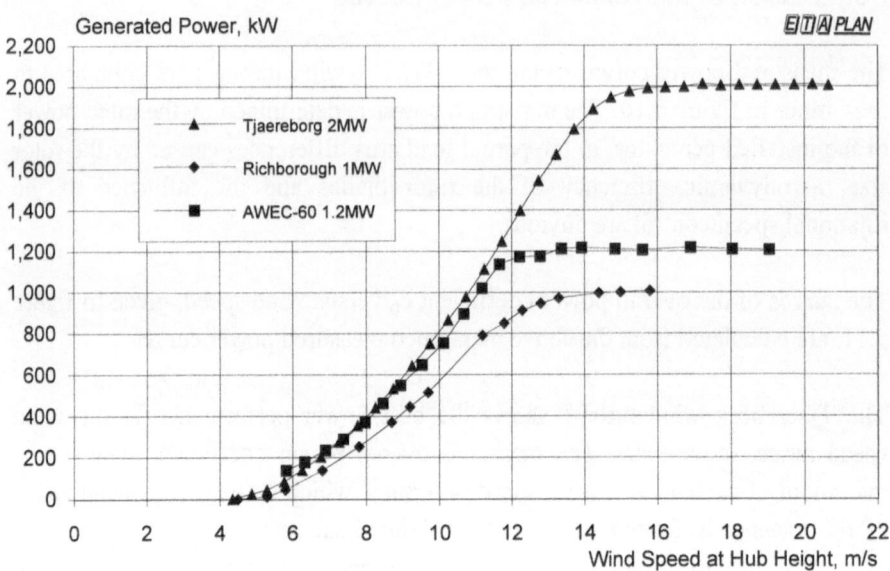

Fig. 5.10 Comparison of the Measured Power Curves of the WEGA Wind Turbines

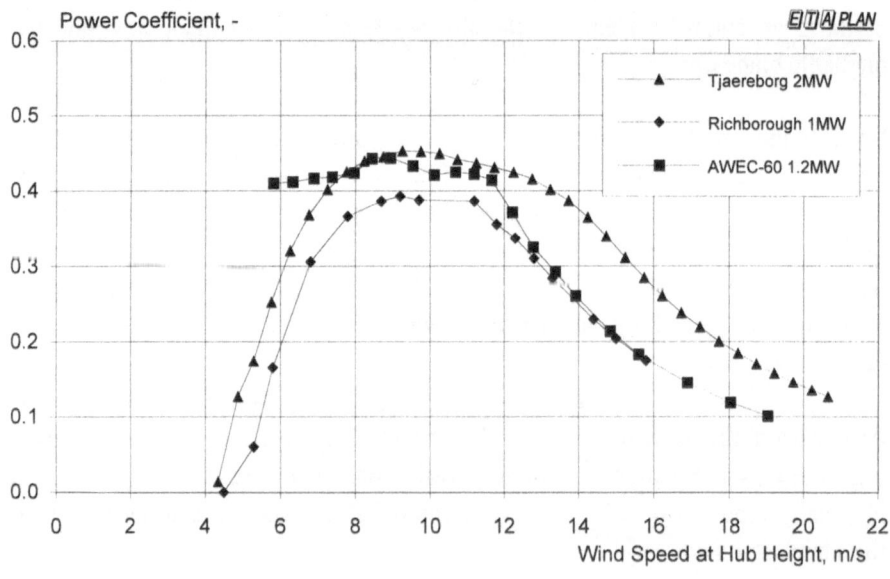

Fig. 5.11 Comparison of the Power Coefficients Versus Wind Speed Calculated from the Measured Power Curves of the WEGA Wind Turbines

5.3 Noise Emission

The acoustic noise emitted by wind turbines can be a serious obstacle for the acceptance of wind energy projects. This is especially the case in areas with a high population density. For this reason the acoustic noise emission of wind turbines is an important item for the licensing authorities - who want to predict the incident noise at a certain distance from the machine - and also for the manufacturer of the turbines - who will have to minimise the noise emission.

The acoustic emissions are composed of a mechanical and an aerodynamical component. Analysis show that for turbines with rotor diameters up to 20m the mechanical component dominates, whereas for larger rotors the aerodynamical component, which is a function of the tip speed, is decisive.

Against this background the contract between the WEGA-machine operators and the Commission stipulated a design target for the incident noise level at a specific distance from the wind turbines:

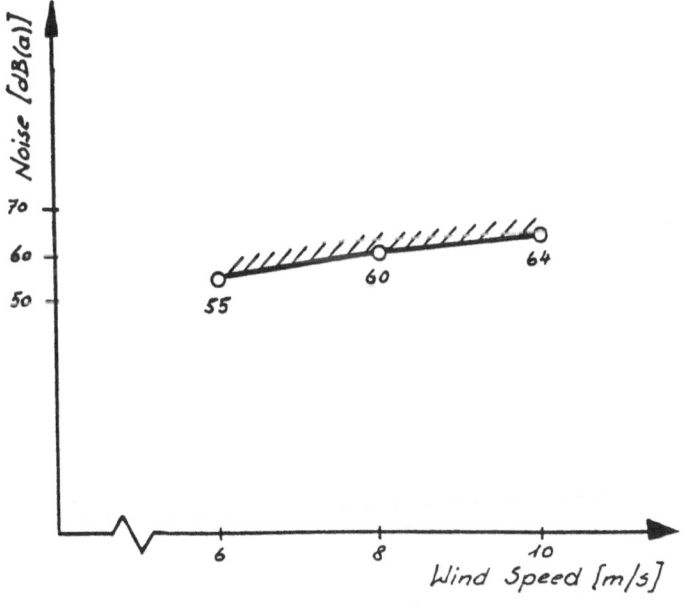

| Noise [db(A)]: | A-weighted equivalent continuous sound pressure level 90m downwind from the wind turbine and at a height of 2m above the terrain |
| Wind Speed [m/s]: | 10 minutes mean wind speed at 10 m height, corrected from the power curve |

Fig.5.12 Design Target in the CEC-Contracts for the WEGA Wind Turbines

The measuring position should be in 90m downwind from the wind turbine at a height of 2.0m above the terrain ("free" surface provided). Measurements in the above mentioned position should be carried out in accordance with the IEA-publication "Recommended Practice for Wind Turbine Testing No.4: Acoustics. Measurements of Noise from Wind Energy Converting Systems" ,1st Edition, 1984 /4/. Furthermore a corrected wind speed at a height of 10m, calculated from the power curve, should be used for the determination of the noise measurements.

Tjaereborg Wind Turbine

Intensive noise measurements have been carried out for the Tjaereborg wind turbine. The measurement procedure followed the improved IEA recommendations, 2nd edition, 1988 /5/.

In accordance with the conditions of the WEGA design target, the table below shows the sound pressure levels as a function of wind speed at a height of 2m above ground level and at a distance of 90m downwind the turbine.

Wind Speed [m/s]	Sound Pressure Level [dB(A)]
4.9	55
5.1	55
5.3	56
6.0	56
6.1	57
7.3	61
8.2	61
8.4	61
8.7	62
9.2	65
10.1	64
10.4	65
11.3	65
11.3	64
11.3	63
13.3	62

Fig.5.13 Two-Minute Average Values of the Wind Speed at a Height of 10m and the Sound Pressure Level 90m Downwind the Wind Turbine at a Height of 2m.

Pure tones in the noise from the wind turbine have not been measured at lower wind speeds. At high wind speeds, i.e. more than 10m/s at hub height, there are two low frequency tones, 143Hz and 197Hz. 197Hz corresponds to the mesh frequency for the second step in the planet gear, while the reason for the tone at 143Hz has not yet been found. However, compared to the dominating part of the noise spectrum, the tones are not clearly audible.

Richborough Wind Turbine

A noise monitoring station was set up 2m above the ground at a distance of 50m from the wind turbine, a 90m location as required in the contract, was not available. Wind speed was measured at a height of 10m.

Figure 5.14a contains three minute traces showing machine noise, normal ambient conditions and a very quiet night on the same trace. The mean wind speed during this measurements is assumed to be 6 to 8m/s. The blade passing frequency can be clearly seen on the upper trace. The mean sound pressure level of the machine in operation is approximately 64dB(A) at the above mentioned measuring position. This is shown as being 20dB(A) above normal ambient at 50m, and 25dB(A) above quiet ambient conditions. For comparison the sound pressure level peaks of a bird song are indicated on the normal ambient trace. An analysis of the spectrum indicates that the generated noise is centred around 250Hz. However, any pure tones are not clearly audible at a height of 2m above ground level.

Figure 5.14b is of interest as it shows the shutdown of the machine. The noise level momentarily peaks at 10dB(A) above normal when the aerodynamic brakes are deployed. The noise then tails away until the disk brakes are applied which generates a loud tonal squeal (which is at 630Hz) of about 10 seconds duration. The sound level drops to its background level 24dB(A) below operating level.

The correlation of the noise level with the wind speed is given in figure 5.14c. The two distinct concentrations of data show clearly the ambient noise level of the area and the noise level of the turbine, which does not increase as much as expected with the wind speed. The resulting mean source power levels from the wind turbine at certain wind speeds in the range 4 to 10m/s are indicated additionally.

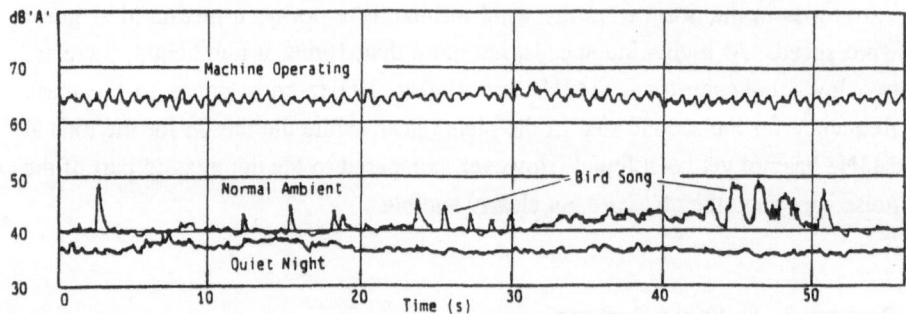

a) Comparison of Different Noise Levels

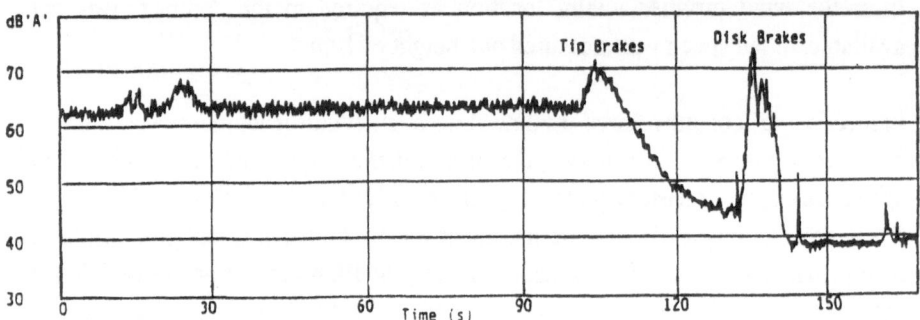

b) Shut-Down of the Wind Turbine

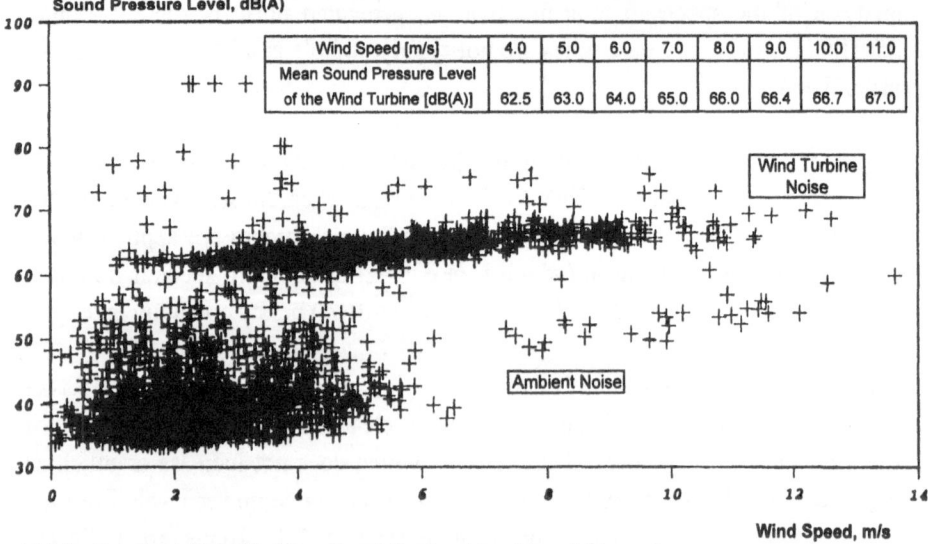

c) Noise Level Against Wind Speed and the Resulting Mean Values

Fig.5.14 Noise Recordings taken 50m from the Richborough Wind Turbine at a Height of 2m

AWEC-60 Wind Turbine

The noise measurements have been performed following the IEA recommendations, 2nd edition, 1988 /5/. The measurement values have been taken at a distance of 76m downwind the turbine and are indicated in figure 5.15.

Wind Speed [m/s]	Sound Pressure Level [dB(A)]
5.20	49.3
6.70	47.6
6.70	51.2
6.90	48.3
10.80	57.4
11.00	60.1
11.20	58.2
11.70	58.3

Fig.5.15 Average Values of the Wind Speed at a Height of 10m and the Sound Pressure Level 76m Downwind the Wind Turbine.

Additional measurements were carried out to determine the influence of the power output of the AWEC-60 wind turbine on its sound emission at approx. nominal wind speed (12m/s). Figure 5.16 shows that there is no significant relation between the machine's power output and the sound power level. Even the small variation of the mean wind speed is the more important controlling factor on the measured sound emission.

Power Output [kW]	Wind Speed [m/s]	Sound Pressure Level [dB(A)]
250	11.7	53.9
500	12.1	57.4
750	11.8	55.9
1000	12.4	58.5

Fig.5.16 Influence of the AWEC-60 Power Output on the Measured Sound Pressure Level

Sound Pressure Levels - Incident Noise

The WEGA design target refers to a measurement distance of 90m downwind the wind turbine.The noise measurements have been actually carried out at this position only for the Tjaereborg machine. The Richborough values, measured at 50m from the wind turbine, and the AWEC-60 measurements at 76m from the turbine need a correction for distance according to the following formulae /5/:

$$L_n = L_i + 20 \lg(R_i/R_n)$$

with: R_i: diagonal distance between measurement point and hub centre

L_i: measured sound power level at a distance R_i from hub centre

R_n: diagonal distance from hub centre to a point 90m from the wind turbine's base

L_n: comparable sound power level at a distance R_n from hub centre

This results in a correction of -3.5dB for the measured mean sound pressure level from the Richborough wind turbine (64dB(A) - 3.5dB = 61.5dB(A)) and -1.1dB for the AWEC-60 measurements (Fig.5.15).

In figure 5.17 the measured and corrected sound pressure levels from the WEGA machines are compared to the CEC design target. Some values from the German wind turbine WKA-60, located at Heligoland - the sister model of the AWEC-60 - are added /6/.

The Tjaereborg measurements show an incident noise level slightly above the design target. The measured values can be seen as typical for a wind turbine with a large rotor diameter and fixed rotor speed. However, as the uncertainty of the sound pressure level measurement is normally estimated to be approximately ±2dB, only three of the measurements seem to be significantly (>2dB) above the design target. Furthermore, the measurements indicate a tendency for the noise emission to saturate at high wind speeds.

Although the rotor diameter of the Richborough turbine is only 55m compared to the Tjaereborg's 61m, an even higher sound pressure level was measured at wind speeds below 8m/s. The characteristics of the partial span controlled rotor, i.e. the slot between the movable tip and the fixed blade, high relative thickness of the selected airfoils even in the outer blade area and the truncated trailing edge in the inner blade section, may all be reasons for this. Furthermore, mechanical noise emissions from the drivetrain may be audible especially at low-wind conditions, because no noise dampening, heavy, materials are used for the lightweight nacelle fairing. However, at wind speeds above 9m/s the sound pressure levels of the Richborough and Tjaereborg wind turbines are in the same range.

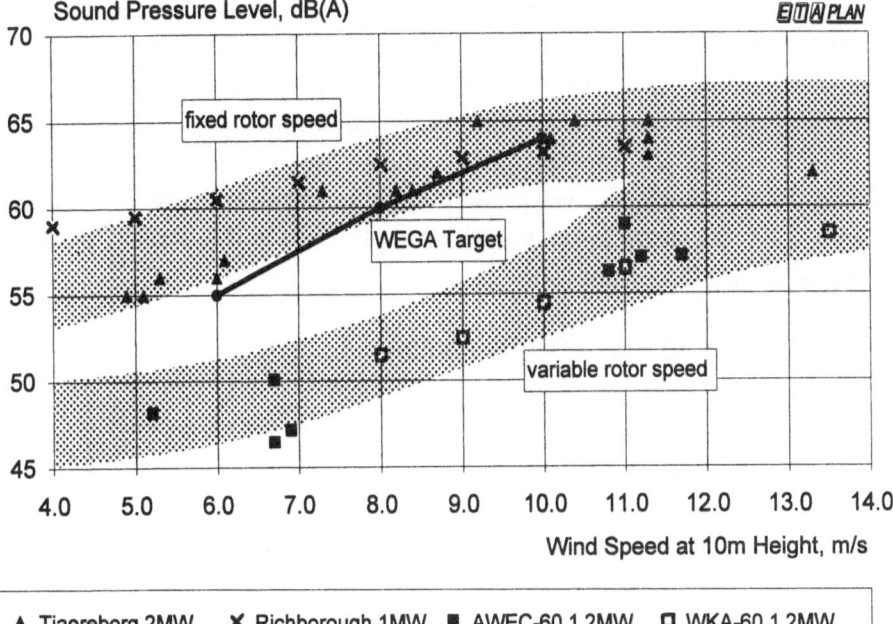

Fig.5.17 Sound Pressure Levels of the WEGA Machines Compared to the CEC WEGA Design Target. The Values from the Richborough and AWEC-60 Measurements are Corrected for the Required Distance of 90m Downwind the Wind Turbine.

Due to its variable rotor speed, giving reduced rotor tip speeds in low-wind conditions, the AWEC-60 is less noisy than the Tjaereborg and the Richborough machine. The values from the WKA-60 turbine -variable speed also - show the same tendency. All measurements are between 5dB and 10dB below the WEGA design target.

The incident noise levels caused by the variable speed machines is shown to be significantly below the noise of the fixed speed turbines at lower wind speeds. At higher wind speeds the variable speed AWEC-60 and WKA-60 both reach their nominal rotor speed and the separation of the two sound pressure ranges seems to diminish.

Source Power Levels - Noise Emission

The source power level represents the noise emission of the wind turbines. A mean wind speed of 8m/s is assumed to be a good reference value for the comparison of the noise emission of different wind turbines because the machines usually are most audible at this wind speed level. At lower wind speeds the noise emission of the turbines is significantly lower. At higher wind speeds the increasing background noise will cover the noise emissions.

Based on the measured sound pressure levels L_i (incident noise) according to figure 5.17, the acoustic source power levels L_W for the machines can be calculated using the relation /5/:

$$L_W = L_i + 10 \lg(4\pi R_i^2)dB(A) - 3dB(A)$$

with: R_i: diagonal distance between measurement point and hub centre

L_i: sound pressure level at a distance R_i from hub centre

and at 2m height above ground level (incident noise)

L_W: A-weighted source power level of the wind turbine (emission)

The measured and corrected sound power levels L_i for the machines have to be adjusted to an average wind speed of 8m/s. Following figure 5.17 the values for the machines are found to be 62.0dB(A) for the Tjaereborg wind turbine, 62.5dB(A) for the Richborough turbine and 51.0dB(A) for the AWEC-60 machine at a wind speed of 8m/s. This leads to the calculated machine's source power levels L_W:

	Incident Noise: Sound Pressure Level Li at 90m Downwind [dB(A)]	Emission: Source Power Level Lw Calculated from Li [dB(A)]
Tjaereborg 2MW	62.0	110.2
Richborough 1MW	62.5	110.5
AWEC-60 1.2MW	51.0	99.1

Fig.5.18 Sound Pressure Levels and Acoustic Source Power Levels of the WEGA Machines at a Wind Speed of 8m/s

At a wind speed of 8m/s the source power level from the variable speed AWEC-60 is calculated to be at approx. 11dB below the emission from the fixed speed machines Tjaereborg and Richborough.

The source power levels of the WEGA turbines are shown in figure 5.19 against the background of a selection of existing wind turbines. The noise emissions of the fixed speed rotors of Tjaereborg and Richborough fit the expected average line. The variable speed concept of AWEC-60 shows a significant lower noise level compared to the today's state of the art performance.

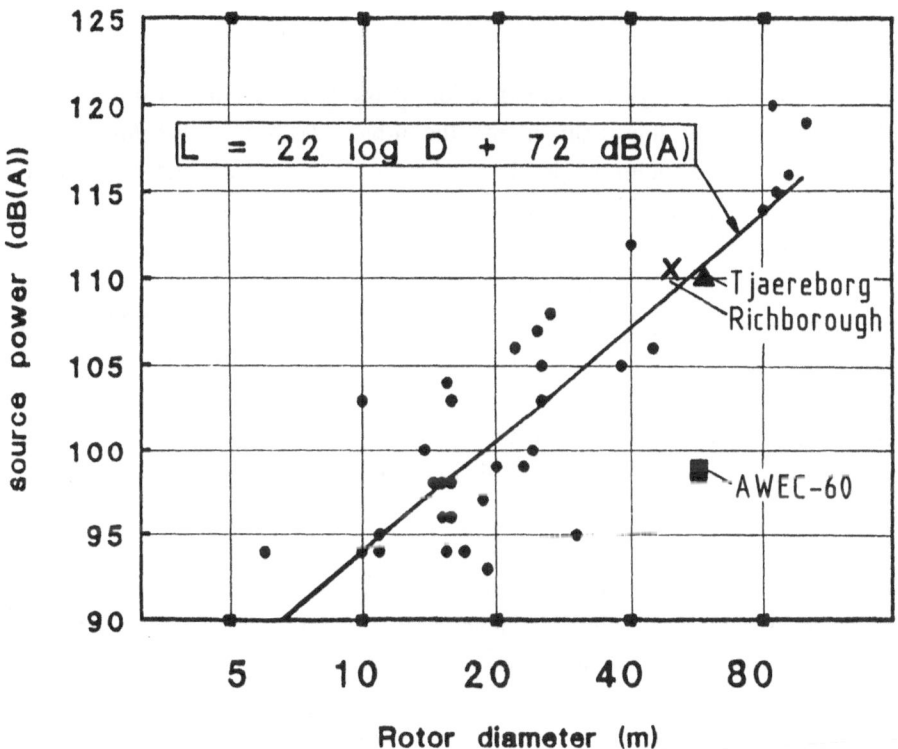

Fig.5.19 Acoustic Source Power Levels of wind turbines as a function of the rotor diameter /7/

5.4   Operational Statistics

Tjaereborg Wind Turbine

The Tjaereborg wind turbine was been synchronised to the grid on March 23, 1988 for the first time. During the commissioning phase and test operation an unexpected harmonic was detected in the power output. The probable cause of this oscillation was found to be the first epicyclic stage in the gearbox where the outer annular ring was too flexible due to lack of support from the gearbox casing. On August 3, 1989 the annular ring broke under full load conditions and the vibration supervision of the gearbox initiated a normal stop which was completed without incidents.

The subsequent redesign and modification of the gearbox took 10 months and after a 6-week recommissioning period, the turbine was back on the grid in May 1990. Due to some additional problems with the bearings of the high speed shaft in the gearbox the regular operation of the turbine did not start until November 13, 1990.

Figure 5.20 shows that the mean availability in the years 1991 and 1992 has been approximately 77%. The monthly percentage values of operating hours and availability for 1992 are given in figures 5.21. The turbine achieved an availability above 90% in its best months. When considering the R&D character of the project this value is unexpectedly high and proves that wind turbines in the multi megawatt class can be reliable.

The operational hours, from the beginning of 1988 until November 30, 1992, are indicated in figure 5.22. The corresponding energy productions - gross per year and net per year - are given in figure 5.23.

From the cumulative totals it can be seen that the Tjaereborg wind turbine produced more than 7GWh of net energy in over 11,200 hours up until the end of November 1992.

For the years 1991 and 1992, during which the turbine has been in regular operation for most of the time, a gross energy production of 3.15GWh/year was achieved. This is consistent with a capacity factor of 18%.

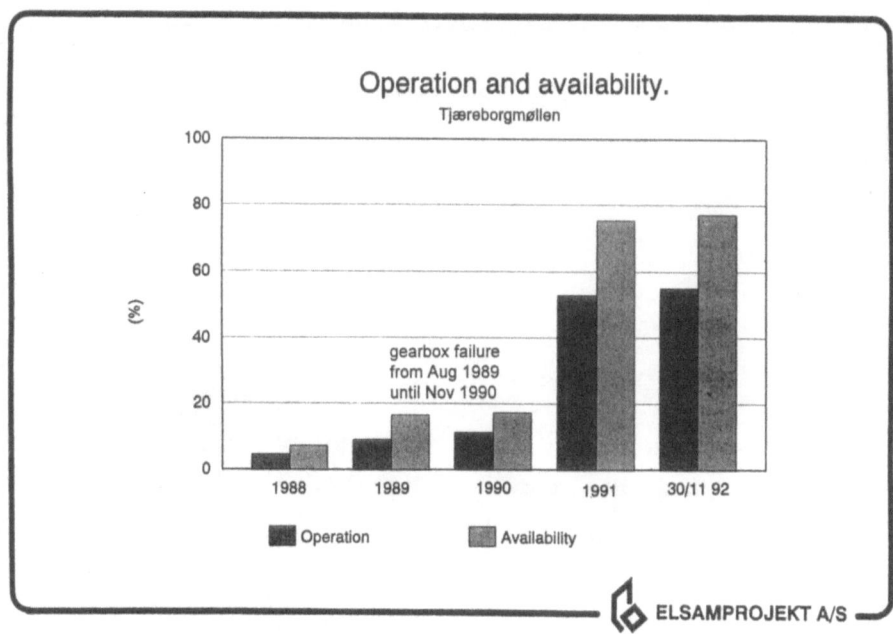

Fig.5.20 Operation and Availability in Percent of Yearly Hours

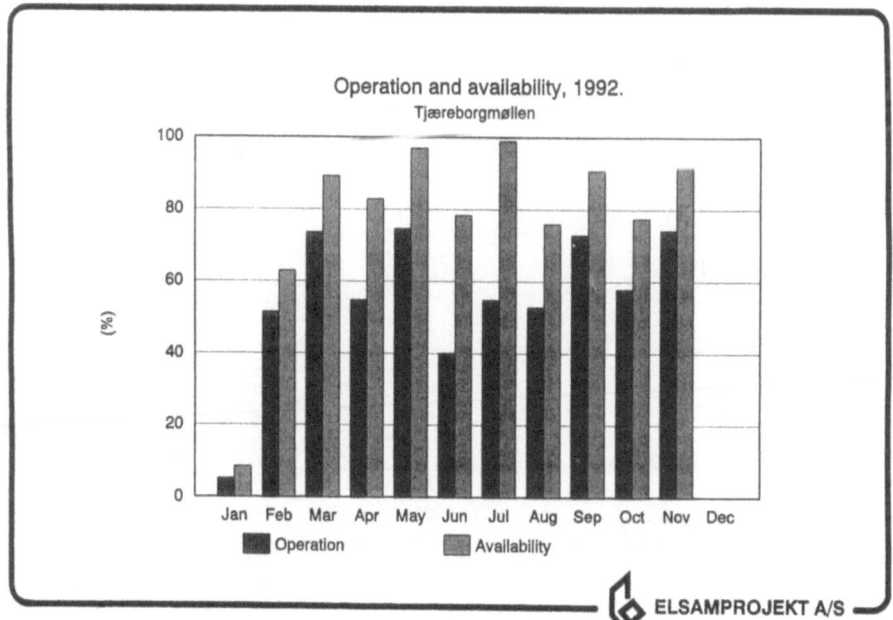

Fig.5.21 Operation and Availability in Percent of Monthly Hours

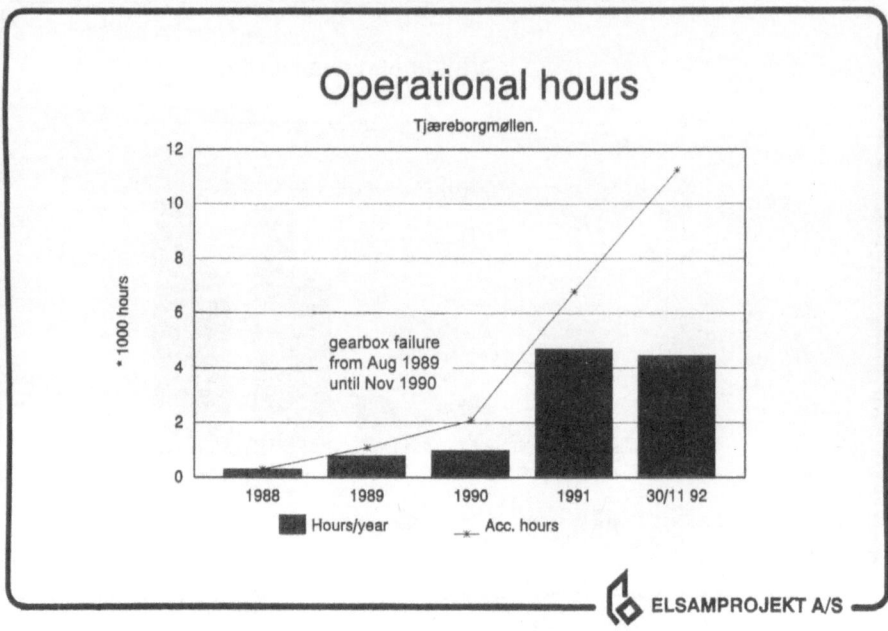

Fig.5.22 Operational Hours of the Tjaereborg Wind Turbine

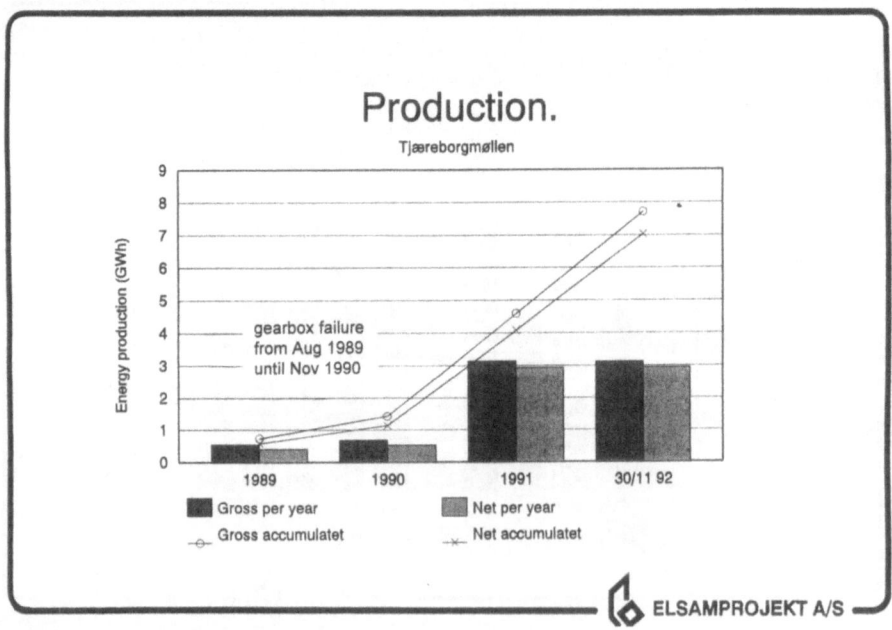

Fig.5.23 Energy Production of the Tjaereborg Wind Turbine

Richborough Wind Turbine

The operational history of the Richborough wind turbine is indicated in figure 5.24. Because of some problems in the commissioning phase, normal operation started in mid of 1990. From then until the end of 1991 good machine performance resulted in increasing availability and a total of 1,712 operational hours were accumulated.

In December 1991 the induction generator suffered two simultaneous phase to earth faults. It had to be completely rewound and was reinstalled in October 1992. On returning to service on October 27, 1992 after an eleven month outage, the level of vibration and noise from the yaw control was considered excessive. The machine's power controller had to be reset to 600kW maximum output. But it is possible to operate for short periods at higher power, for the purposes of the measurement programme.

Figure 5.25 shows the energy production of the Richborough turbine. From July 1990 until December 1991 the machine produced at least 1,482MWh. This gives a mean capacity factor of 12% for this period. It is estimated that the machine has been available for approximately 7,000 hours up to now.

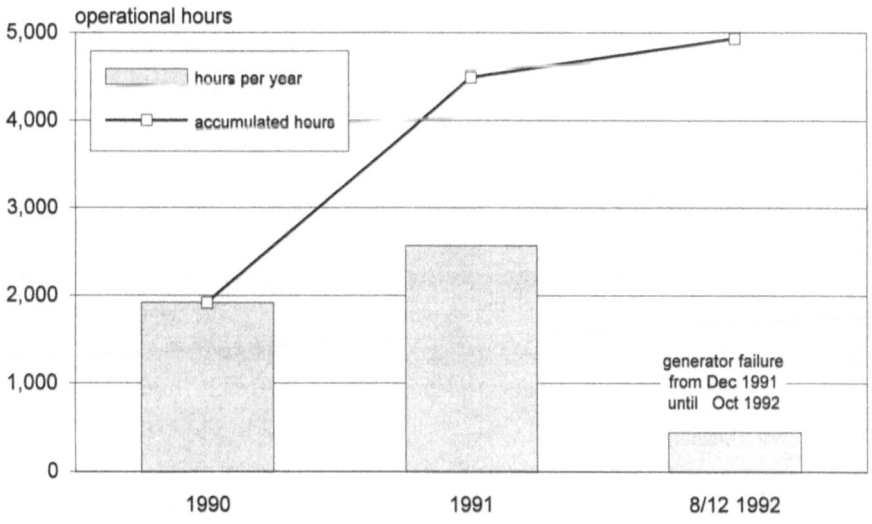

Fig.5.24 Operational Hours of the Richborough Wind Turbine

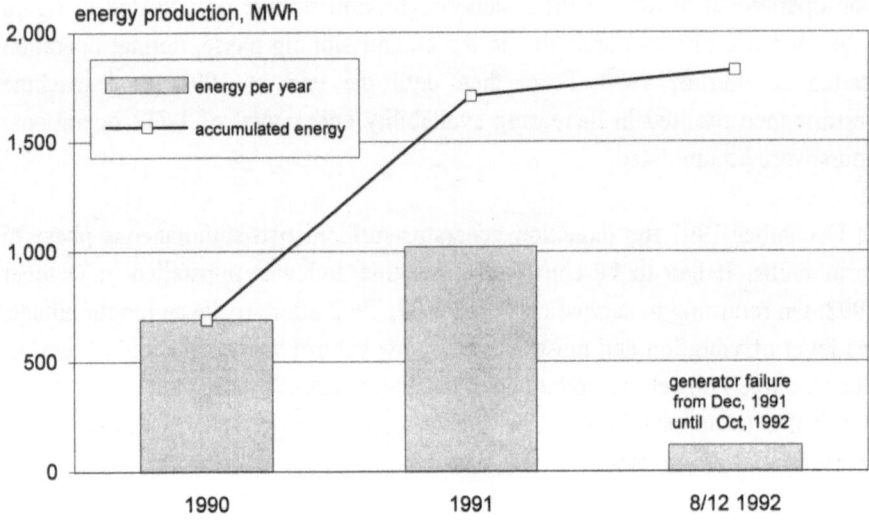

Fig.5.25 Energy Production of the Richborough Wind Turbine

AWEC-60 Wind Turbine

The AWEC-60 turbine had to overcome some problems occurring in the yaw system and the pitch control system during the early operational period (1990). The result of this was that the availability did not achieve a satisfactory level.

In May 1991 a severe crack in the load carrying structure of one rotor blade was detected. The blade had to be dismounted from the turbine. The carefully planned repair procedure was followed by eigenfrequency tests and strain measurements. At the end of September 1991 the blade was remounted and the wind turbine was commissioned again but the maximum power was limited to 300kW. This is the reason for the low energy production during the relatively high number of operational hours achieved in 1992 (Fig.5.27 and Fig.5.28).

From January 1991 until April 1991 the AWEC-60 wind turbine achieved its best level performance to date. Based on an availability of 40% the machine produced 301MWh in these four months. The corresponding capacity factor is calculated to be slightly above 10%.

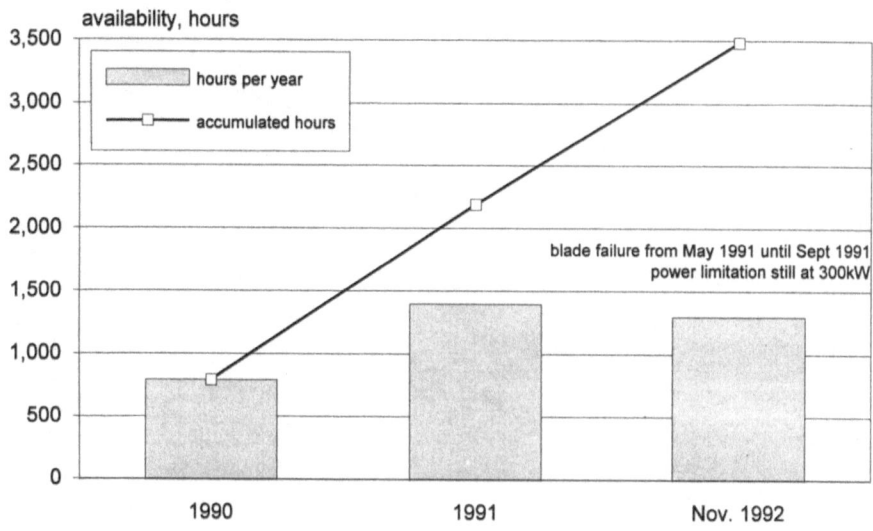

Fig.5.26 Availability of the AWEC-60 Wind Turbine

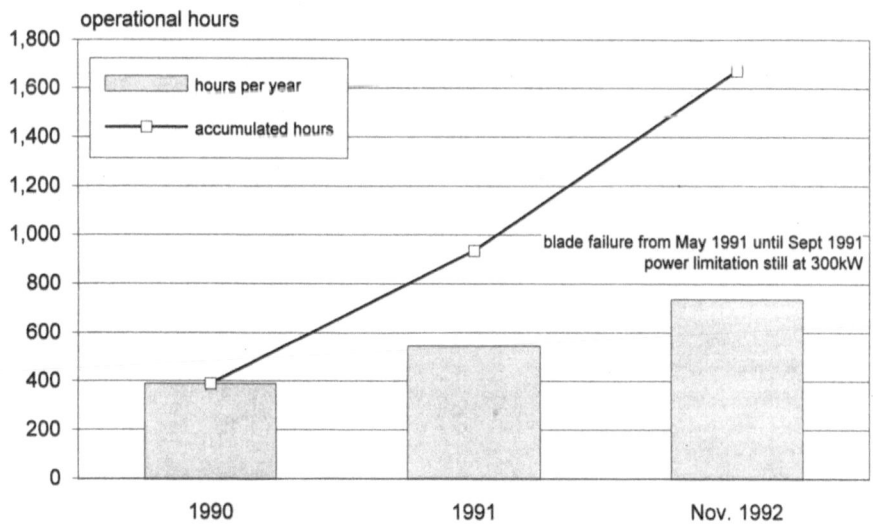

Fig.5.27 Operational Hours of the AWEC-60 Wind Turbine

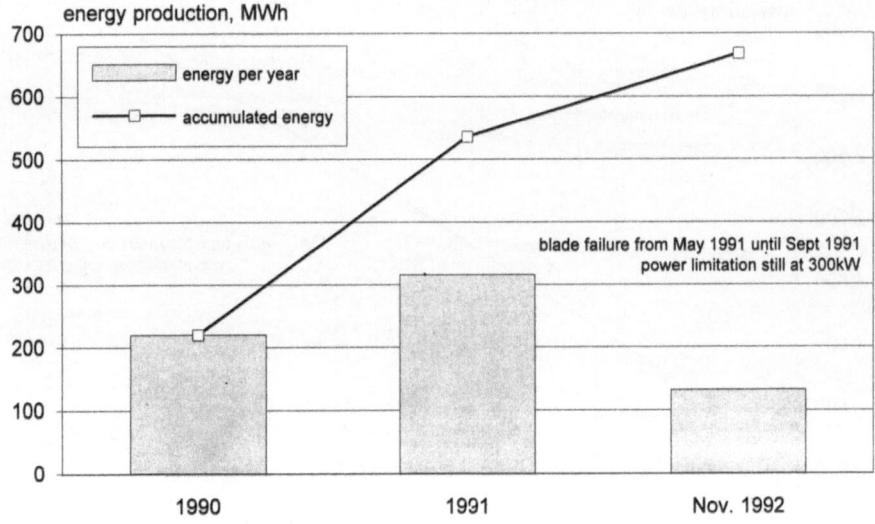

Fig.5.28 Energy Production of the AWEC-60 Wind Turbine

6
Future Outlook: WEGA II Programme

Although the WEGA programme has not yet come to an end one important conclusion can already be drawn. Large wind turbines cannot become commercially viable using the technology upon which the current WEGA machines are based. This is clearly true for the Tjaereborg wind turbine and the AWEC-60 if less so for the Howden Richborough machine. The main reason lies in the high, uncompetitive costs of the towerhead machinery which, in turn, are due to the high component weights.

In the commercial wind turbine industry, on the other hand, a continuous growth in the size of series produced commercial machines can be observed. These machines are becoming more lightweight in construction and, also, have gradually increased in size up to the 500kW; the level which is the industry standard today. The technological gap between the commercially driven developments of the small to medium sized machines and R+D concepts of the large experimental wind turbines is obvious and in need of investigation.

In order to address this problem DG XII of the CEC initiated a strategic study entitled - **Study on the Next Generation of Large Wind Turbines**. The study is being carried out by an international group of experts with the aims of identifying the optimum size range of wind turbines and also of comparing the cost effectiveness of different wind turbine concepts including commercial and R+D designs. The cost calculations include all factors in a turn-key installation together with operation and maintenance costs over the lifetime of the plant. The results of the study should assist with the next step in the development of large wind turbines.

The main economic conclusions of the strategic study are that there is no clear optimum size in the size range with rotor diameters from 30 to 80m. Unit costs of electricity production slowly increase with size but the increase is so slow that it can be offset by improvements in design and by economies of scale in other cost areas such as site and operation costs. Interim results from the study have also shown that energy generation costs from large wind turbines could be reduced to the same levels as today's commercial wind turbines if lightweight design principles are adopted.

While economics is usually the first consideration with any new technology it is likely that the future potential of wind utilisation will also be influenced, in an important way, by the environmental impact of the wind turbines.

Within the next few years, it can be expected that, the economic case for wind energy exploitation, in areas with good wind regimes, will become clearly established. However, when environmental restrictions are taken fully into account it is likely that the public will accept wind energy developments on limited land areas in specific locations only. In this situation the over-riding consideration may be to maximise the energy production from such sites rather than to obtain absolute minimum costs.

It is a matter of fact that large turbines offer considerably more power and energy for a given land area. In all cases this is due to the higher hub height but also in the particular situation where they are installed in a linear arrangement an additional advantage is that more rotor swept area can be achieved with large machines on the same stretch. In the future, if the average size of wind turbines is in the range of 50-80m rotor diameter instead of the today's 20-30m, the installable power as well as the energy yield will be between 1.5 to 2.5 times higher for a restricted set of land areas and locations.

The noise emission of a wind turbine is one of the major factors with respect to its environmental acceptability. It is directly related to the availability of land. The quieter the wind turbines are the more land that can be used for wind energy developments. By far the most important factor is the blade tip speed.

'Low noise' concepts are very important for large rotors, particularly for two-bladed machines with blade tip speeds rated at around 100m/s. The capability of reducing the rotor speed at lower wind speeds is essential. It must be expected that variable speed or two speed electrical systems will be important design and operational elements for large two-bladed rotors in the future in order to reduce noise.

The visual impact of turbines on the landscape is a real factor in the environmental planning, even though the impact cannot be judged objectively. Larger turbines are visible from greater distances, but this has to be set against the large number of small turbines which is needed to give the same level of output.

Encouraged by the results of the strategic study described above the European Commission launched a call for tender for the development of a next generation of large wind turbines. Setting a deadline of February 14, 1992 the Commission asked for technical proposals, which, if accepted, could be financially supported in the framework of the second R&D programme Joule II.

The technical specification in the call for tender referred to large wind turbines with a rotor diameter of at least 45m and a rated power output around one megawatt. Furthermore, a variety of technical criteria had to be satisfied in the proposals (Fig.6.1). The Commission has used these criteria to form a judgement of the economic potential of each proposal. The main policy has been to support only technical concepts which clearly have a real chance to achieve economic viability, but also a variety of different technical approaches will be included in the overall R&D programme.

Required Size	rotor diameter: ≥ 45m
	rated power in the range of 1MW
	energy production at reference site: ≥ 1.8GWh/year
Reference Site	annual mean wind speed 7.5m/s at 50m height
Wind Regime	wind shear exponent 1/7
	Weibull distribution with k=2.0
Lifetime	20 years (of 100% availability)
Noise Target	source power level 'less than 22logD+72' dB(A)
	(D = rotor diameter)
Power Quality	according to IEA recommended practices
Electromagnetic	
Compatibility	directive 89/336/EEC
Safety Provisions	in line with IEC/TC 88 (draft)

Fig.6.1 Technical Requirements of the Call for Tender for Large Wind Turbine Developments Under the New WEGA II Programme

From 10 eligible proposals received, seven were accepted for the WEGA II programme (Fig.6.2). The total support given to the above projects will be 7.0MioECU. Its relationship to the support for generic wind-energy R&D (4.0MioECU) indicates clearly the Commission's priorities.

Main Proposer	Partners	Grant kECU	Rotor Diam.	Rated Power	No. of blades	Innovative elements
Nordic (SE)	LM (DK) Flender (DE) Vattenfall (SE)	400[1]	52.0	1.0	2	combination of: stall & var. speed, teetered hub, soft structure
Enercon (DE)	Aerpac (NL)	1,400	55.0	1.0	3	no gearbox, new generator-type, individual blade pitch control
Nedwind (NL)[2]	Garrad H. (UK)	250	52.6	1.0	2	optimisation of stall characteristics by means of additional pitch control, six generators,
WEG (UK)	Pfleiderer (DE)	1,200	50.0	1.0	2	combination of: teetered hub, variable speed, electric pitch control
Heidelberg Motor (DE)	Union Fenosa (SP) RWE (DE) WIP (DE) NTUA (GR)	1,600	58.0	1.2	2	vertical axis rotor, no gearbox, new generator-type,
Bonus (DK)	Risø (DK) Garrad (UK) TUD (DK) LM Glasfiber (DK) AN Maschinenbau (DE) Elsamprojekt (DK) Flender (DE)	1,300	50.0	0.75	3	advanced compact drive train, optim. stall blades, noise reduction elements
Vestas (DK)[3]	Siemens (DE) Hanssen (BE)	500	50.0	1.0	3	compact nacelle, adv. rotor blades new pitch control

[1]   EC-partner costs only,
[2]   Measurement phase only to be supported.
[3]   Design phase to be supported in this programme.

Fig.6.2 Proposals to be Supported in the Large Wind Turbine Programme WEGA II, /8/

As figure 6.2 shows, the WEGA II programme compares the 'best examples' of several concepts such as stall control and pitch control, 2 and 3 blades, vertical and horizontal axis etc. All the horizontal axis proposals are lightweight, seen as the key to low cost. Two machines have an advanced electrical conversion system without the need of a mechanical step-up gearbox. The WEGA II programme will take the exploration of the relative advantages of different technical options a step further.

It should also be remembered, that WEGA II does not include all of the large wind turbines being developed in Europe. Other important projects have been proposed for support under the CEC DG XII's Thermie programme. They have smaller R&D content. Also other national programmes continue with large machine developments.

A variety of large wind turbines in the megawatt range will be developed and tested during the next three years. It can be expected that at least some of the prototypes will establish the technical basis for an economically viable megawatt wind turbine. Certainly substantial additional work will be necessary to put these machines into series production, but as of today there are good prospects that large wind turbines, in the megawatt power range, will eventually achieve a commercial application in the late nineties. The utilisation of wind energy will then enter into a new phase of importance.

References

WEGA Project Documentation
(not indicated specifically when they are quoted)

Tjaereborg 2MW Wind Turbine

-   **ELSAMPROJECT A/S:** *The Tjaereborg Wind Turbine - Final Design Report Volume 1&2.*
    CEC DGXII Contract EN3W.0048.DK, September 1992
-   **ELSAMPROJECT A/S:** *Tjaereborg 1 - 60m/2MW Wind Turbine - As Built Report Volume 1 & 2.*
    CEC DGXII Contract EN3W.0048.DK, July 1990
-   **ELSAMPROJECT A/S:** *The Tjaereborg Wind Turbine - Final Report.*
    CEC DGXII Contract EN3W.0048.DK, September 1992

-   **ELSAMPROJECT A/S:** *The Tjaereborg Wind Turbine, Power Curve Measurements, November 1990 - August 1991.*
    CEC DGXII Contract EN3W.0048.DK, November 1991
-   **ELSAMPROJECT A/S:** *The Tjaereborg Wind Turbine, Noise Measurements on the Tjaereborg Wind Turbine, CEC-Contract Measurements.*
    CEC DGXII Contract EN3W.0048.DK, November 1991
-   **ELSAMPROJECT A/S:** *The Tjaereborg Wind Turbine, Site Characteristics.*
    CEC DGXII Contract JOUR-0025-DK (MB), February, 1992
-   **ELSAMPROJECT A/S:** *The Tjaereborg Wind Turbine, Quarterly Reports*
    CEC DGXII Contract EN3W.0048.DK

- **Friis, P.:** *Tjaereborgmollen at Esbjerg - The Danish 2MW Wind Turbine.*
  Proceedings of the Euroforum - New Energies Congress, Saarbrücken, 1988
- **Friis, P.; Christiansen, P.S.; Hansen, K.S.:** *Operating Experience with the Tjaereborg Wind Turbine, 1989 - 1990.*
  European Community Wind Energy Conference, Madrid, 1990
- **Friis, P.;** *Personal Information.*
- **Christiansen, P.S.;** *Personal Information.*

Richborough 1MW Wind Turbine

- **Central Electricity Generating Board (CEGB):**
  The Richborough 1MW Wind Turbine - Design Report.
  CEC, DGXII Contract EN3W.0049.UK, January 1988
- **Central Electricity Generating Board (CEGB):**
  The Richborough 1MW Wind Turbine - As Built & Final Report.
  CEC, DGXII Contract EN3W.0049.UK, to be published in 1993

- **James Howden and Company Ltd.:** *The Howden 1000/55 Wind Turbine.*
  December, 1986
- **PowerGen:** *The Richborough Wind Turbine.*
  PowerGen plc Richborough Power Station, Sandwich, 1990
- **PowerGen:** *Richborough Wind Turbine, Quarterly Report, Jan 1991.*
  Technology and Research Department, Ratcliffe-on-Soar, Nottingham, March 1991
- **PowerGen:** *Richborough Aerogenerator Noise Measurements.*
  Technology and Research Department, Ratcliffe-on-Soar, Nottingham, March 1991

- **Milborrow, D.J.; Bedford, L.A.W.; Passey, D.J.:**
  The 1MW Wind Turbine at Richborough.
  Proceedings of the Euroforum - New Energies Congress, Saarbrücken, 1988
- **Rea, J.; Boston, A.:** *The Commissioning and Early Operation of the 1MW Wind Turbine at Richborough.*
  Ratcliffe Technology Centre PowerGen plc, United Kingdom, 1990
- **Pearce, D.L.:** The *Richborough H55 1MW Wind Turbine.*
  Report to WEGA Project Meeting, Sardinia, June 1992
- **Boston, A.:** *The Richborough 1MW Wind Turbine.*
  Report for WEGA Project Status Meeting, Fredericia, October 1992

- **Moreton, W.:** *Howden HWP 1000/55 at Richborough, Kent.*
  Status for WEGA Design Review and Evaluation Project,
  Renewable Energy Systems Ltd., December 1992
- **Pearce, D.L.; Hilliar, R.A.:** *Richborough H55 1MW Wind Turbine.*
  Report to WEGA Project Meeting, Greenford, December 1992
- **Pearce, D.L.:** *Personal Information.*

AWEC-60 1.2MW Wind Turbine

- **UNION FENOSA, CIEMAT-IER, ASINEL, MAN:**
  *AWEC-60 Advanced Wind Energy Converter in the One-Megawatt-Class -
  Final Design Document.*
  CEC, DGXII Contract EN3W.0047.E, Madrid, March 1987
- **UNION FENOSA, CIEMAT-IER, ASINEL, MAN:**
  *AWEC-60 Advanced Wind Energy Converter in the One-Megawatt-Class -
  Final Document.*
  CEC, DGXII Contract EN3W.0047.E, Madrid, December 1991

- **Union Fenosa, Ciemat-IER:** *Status Report for the AWEC-60 Wind Turbine*
  La Coruna, October 1991
- **Union Fenosa:** *Experiences from the AWEC-60.*
  Madrid, November 1992
- **Union Fenosa, Ciemat-IER:** *Quarterly Reports on Operation and
  Production.*
  CEC, DGXII Contract EN3W.0047.E

- **Avia, F.:** *Study of Wind Conditions in Cabo Villano.*
  CIEMAT IER, Madrid, June 1987
- **Caño, A.:** *Design and Construction of a 1.2MW/60m Wind Turbine -
  Project AWEC-60 - Plant to be installed in Cabo Villano (La Coruna, NW
  Spain).*
  Proceedings of the Euroforum - New Energies Congress, Saarbrücken, 1988
- **Soria, E.; Matas, A.:** *Operation and Maintenance Experiences and Test
  Results from the AWEC-60 Project.*
  European Wind Energy Conference, September 1990
- **Avia, F.:** *Power Curve of the AWEC-60 Wind Turbine.*
  CIEMAT IER, Madrid, December 1991

- **Santamaria, M.F.; Piñiero, E.G.-l.:** *Informe de las Medidas de Emesiones Acusticas Procedentes del Aerogenerador AWEC-60.*
  Asinel, Departemento de Electrotecnicia, Mostoles, March 1992
- **Matas, A.:** *Personal Information.*

LS1 3MW Wind Turbine

- **Simpson, P.B.; Lindley, D.; Prosser, E.R.; Walker D.E.:**
  Construction of the LS1, 3MW Wind Turbine.
  European Wind Energy Association Conference, Rome, October, 1986

Gamma 60 1.5MW Wind Turbine

- **Avolio, S.; Calo, C.; Foli, U.; Rubbi, L.; Casale, C.; Sesto, E.:**
  GAMMA 60 1.5MW Wind Turbine Generator.
  European Community Wind Energy Conference, Madrid, September 1990

Further Publications

/1/ **Troen, I.; Petersen, E.L.:** *European Wind Atlas.*
   Risø National Laboratory, Roskilde, Denmark
   Published for the Commission of the European Communities, Directorate-General XII for Science, Research and Development
   Brussels, Belgium, 1989
/2/ **Petersen, E.L.:** *Danish Windatlas.*
   Risø National Laboratory, Roskilde, Denmark, 1981
/3/ **Frandsen, St.; Pedersen, B.-M.:**
   Expert Group Study on Recommended Practices for Wind Turbine Testing and Evaluation - 1. Power Performance Testing.
   Submitted to the Executive Committee of the International Energy Programme for Research and Development on Wind Energy Converting Systems, 2. Edition, 1990

/4/ **Ljunggren, St.; Gustafsson, A.; Trenka, A.R.:**
Expert Group Study on Recommended Practices for Wind Turbine Testing and Evaluation - 4. Acoustics, Measurement of Noise Emission from Wind Energy Converting Systems.
Submitted to the Executive Committee of the International Energy Programme for Research and Development on Wind Energy Converting Systems, 1. Edition, 1984

/5/ **Ljunggren, St.; Gustafsson, A.:**
Recommended Practices for Wind Turbine Testing - 4. Acoustics. Measurement of Noise Emission from Wind Turbines.
2. Edition, 1988

/6/ **Hadler, C.; Götze, H.-J.:**
Windkraftanlage WKA 60/1 Helgoland - Geräuschmessungen.
Germanischer Lloyd, Hamburg, 13.11.1990.

/7/ **Borg, van der, N.J.C.M.; Stam, W.J.:**
Acoustic Noise Measurements on Wind Turbines.
ECWEC-Conference 1989

/8/ **Caratti, G.; Zervos, A.:**
Joule II Programme Wind Energy R&D, Commission of the European Communities, DG XII.
Winddirections, Vol XII No. 1, 1992

/9/ **Hau, E.:** *Windkraftanlagen.*
Springer-Verlag, Berlin, Heidelberg, 1988